THE PHYSICS OF CAPITALISM

The Physics *of* Capitalism

HOW A NEW POLITICAL ECOLOGY
CAN CHANGE THE WORLD

Erald Kolasi

MONTHLY REVIEW PRESS
New York

Copyright © 2025 by Erald Kolasi
All Rights Reserved

Library of Congress Cataloging-in-Publication Data
available from the publisher

ISBN 978-1-68590-090-8 (paperback)
ISBN 978-1-68590-091-5 (cloth)

Typeset in Minion Pro

MONTHLY REVIEW PRESS, NEW YORK
monthlyreview.org

5 4 3 2 1

Contents

Acknowledgments | 7
Introduction: Nature and Society | 9

1. Growth and Scale in Economics | 33
2. The Neoclassical World | 70
3. Theories of Economic Growth and Development | 103
4. The Bionomic Disruption and the Future of Humanity | 146
5. Energetic Conversions in the Economic Process | 176
6. The Ecodynamic Synthesis | 196
7. Technological Dynamics of Growth and Stability | 214
8. Energy and Technology in the Social Sphere | 246
9. The Industrialization of Britain: A Case Study | 271
10. A New Vision | 305
11. The Ecological State | 321
12. The Valerist Cosmopole | 341

Notes | 353
Index | 384

*For my loving wife, Laura,
and my two beautiful children, Logan and Brooklyn*

Acknowledgments

I am deeply grateful to my wife, Laura, for encouraging me to finish this book, giving me time to do so while juggling life with two young children, and always valuing my thoughts and ideas. I am thankful to my father, Bektash, for raising me to be a critical thinker, and for the many incisive debates and discussions we've had on the issues presented in this book. I am also profoundly indebted to my grandmother, Dhurata, who taught history and geography and inspired me to have a deep curiosity about the nature of the world from a very young age.

This book would not have been possible without the contributions of the Monthly Review team. I am grateful to Michael Yates for his engaging analysis of the initial drafts, and to John Bellamy Foster for giving me a chance and accepting my essays for publication in the magazine. Beyond MR, I am thankful to Blair Fix, one of my collaborators on prior projects, for offering me fresh perspectives on the limits of current economic theories. I would also like to acknowledge my PhD advisor, Joseph Weingartner, for being an excellent mentor and professor, and for giving me a renewed appreciation for physics and the natural world.

INTRODUCTION

Nature and Society

Nature is embroiled in politics, often in the most unexpected ways. The Netherlands is the ultimate master of agricultural efficiency; it produces 6 percent of Europe's food with just 1 percent of Europe's farmland.¹ But not everything is rosy with the Dutch. In October 2019, farmers in the Netherlands made headlines around the world when they took to their tractors and started blocking roads and highways throughout the country.² They were outraged by recent court rulings and government proposals designed to reduce nitrogen emissions. The overabundance of nitrogen in Dutch ecosystems was sabotaging the country's biodiversity, leading to the rapid growth of grasses and harmful weeds that could devastate insect and bird populations, in addition to the country's pristine landscape. Even worse, many of the country's shallow water systems had experienced horrible levels of pollution; some urban water areas in the Netherlands started to look like dumping grounds for hazardous waste. The government responded by mandating new standards for limiting the excessive nitrogen pollution being produced by farms and factories. But a Dutch court ruled in May 2019 that the government's plans didn't go far enough and violated European Union regulations; the decision immediately halted some 18,000 construction projects worth roughly 14 billion euros.[3] All of a sudden, farmers could no longer make critical repairs, add more livestock, or expand their operations in any significant way. Their anger finally boiled over in the fall. Three years later, in the summer of 2022, Dutch farmers started protesting again by dumping garbage and manure all over national highways. Where did all this excess nitrogen come from, and why was it such a big deal, to the point where it brought a country like the Netherlands to a screeching halt on multiple occasions?

Modern agriculture relies heavily on artificial fertilizers, which usually contain large amounts of nitrogen and phosphorus. Plants need these critical nutrients to power their growth, especially since nitrogen and phosphorus are both essential components of many proteins and other biomolecules. But the fertilizers don't always stay in the fields where they're dumped; rain and seeping groundwater can carry them to other areas. And that's where the challenges begin. The excessive concentration of nutrients in a particular environment is known as *eutrophication*.[4] When fertilizer waste makes its way into rivers, estuaries, lakes, and oceans, bacteria and other tiny organisms use the nutrients to grow large colonies of algae, sometimes called algae blooms. These blooms block sunlight for aquatic organisms that live below the surface. And once the algae themselves die, the decomposition of their remains sucks out the rest of the oxygen in the water, turning the area into a "dead zone." There are hundreds of dead zones around the world by now, with the largest ones reaching the size of small countries. Together, eutrophication and dead zones can harm fishing stocks, reduce tourism, and disrupt water supplies. In 2014, an algae bloom in Lake Erie disrupted the water supply of 500,000 people in the city of Toledo for several days.[5] Algae blooms can look curiously harmless from a distance, but they're not so innocent when they bring down entire ecosystems.

The effects of nature on human civilization are often gradual and cumulative, but the natural world can sometimes deliver a sudden blow that completely reshapes a particular society. In November 1970, the intense winds and storm surges of Cyclone Bhola struck East Pakistan, a large enclave of Pakistan created after the British partition of India in 1947. The brutal cyclone killed up to 500,000 people, making it one of the deadliest recorded natural disasters in human history.[6] Entire villages were wiped out and large parts of the region were utterly devastated. Public anger boiled over, especially at the government's lethargic response to a massive humanitarian tragedy. The political consequences were felt immediately. In December, the pro-autonomy Awami League won a landslide victory at the polls, but officials in West Pakistan refused to accept the results because they didn't want to hand over the reins of power to a Bengali political party. The rising political instability among the various factions contributed to the outbreak of war in 1971. By the end of the conflict, East Pakistan was gone, and a new nation was born in its place: Bangladesh.

Cyclones and algae blooms can be massive systems stretching over vast distances, but nature can also impact humanity through much smaller biophysical vectors. The tsetse fly lives in sub-Saharan Africa and carries dangerous microbes capable of killing cattle and other livestock in large numbers. The disease produced by these microbes, characterized by fever and anemia, is known as *nagana* when it affects animals and "sleeping sickness" when it affects humans. It's estimated that bites from tsetse flies kill at least three million cows and other livestock in sub-Saharan Africa every year, causing billions of dollars in financial damage along the way.[7] It's bad enough that some farmers have historically avoided large parts of sub-Saharan Africa where agriculture could otherwise thrive, simply so they don't have to deal with tsetse flies. Many people have argued that the deadly flies are a major reason for Africa's economic underdevelopment relative to other continents, though certainly other historical factors have also played a major role. Nagana is far from the only disease to impose devastating effects on humanity. From smallpox and cholera to typhus and Covid, epidemic diseases have always been some of the most decisive drivers of human history. In the twentieth century alone, smallpox may have killed up to 500 million people, about as many as have died in all the wars humanity has ever fought.[8] The eradication of smallpox by the 1970s, following a massive global vaccination campaign, was by far the greatest public health victory in human history.

The complex interactions between nature and human society do not just unfold on the surface of the planet. In many large cities around the world, people extract water from underground water reservoirs called aquifers. But some of these cities are at least partially built on soft soils like clay-rich lake beds, so when too much water is extracted, the soft sheets of land underneath start to compress and undergo a process known as *subsidence*, meaning that the ground literally sinks. Subsidence is a huge problem that's currently affecting many places around the world, especially dominant metropolises like Mexico City and Jakarta. In Mexico City, differential rates of subsidence across various districts have led to broken pipes and roads along with collapsing houses and buildings.[9] Neighborhoods that were once sitting on flat ground now look like they were built on gently rolling hills. In Jakarta, the capital of Indonesia, subsidence and rising sea levels have cleared out numerous coastal neighborhoods and contributed to massive flooding problems. Throw in excessive pollution levels and extreme traffic congestion and you get a situation where it's becoming almost unbearable to live in certain parts of the city. Jakarta's ecological problems have gotten so bad that the Indonesian government has decided to run away and build an entirely new capital, called Nusantara, on the island of Borneo.[10] Indonesian officials claim that the new capital city will be "green" and environmentally sustainable, but its very construction has led to the devastation of pristine forests and other critical habitats, to say nothing of the harmful pollutants and greenhouse gases that are also being released into the atmosphere. This example nevertheless underscores a common pattern that repeats throughout human history: when people make a mess of things in one area, they either migrate to other areas or try to somehow shift the locus of ecological degradation. For example, rich countries are notorious for selling much of their trash to poor countries, in effect exporting some of their ecological problems away by taking advantage of desperate countries that need money or other resources.[11]

Cyclones, epidemics, dead zones, and subsidence are just some of the many ways that the natural world intersects with human civilization. Even though we often think of our social lives as being somehow separate from nature, there's a profound connection between the world "in here" and the one "out there." We sometimes have this sense of the natural world as an obscure force swarming with uneventful distractions. We do not consider ourselves as physical systems guided by natural conditions; we see ourselves as independent agents doing whatever we want. We have a gut-level belief in the power of our infinite will, and the idea that other things can stop us or challenge us seems laughable. If we wish to "tame" nature, we build dams to control rivers. We blow holes through mountains and construct tunnels. We like to brag about the impact that our actions have on nature, by showing off our elegant skyscrapers, gorgeous bridges, and lengthy canals. When we wish to "protect" nature, we establish parks and start exploring them through hiking, climbing, and camping. In this frame of mind, we are *active* and nature is *passive*; we do things and nature just sits there. But if we are truly so detached from nature, there would be no reason to seek shelter from the heavy rain, no reason to flee from a hungry tiger, no reason to consume oxygen from the atmosphere, no reason to take long walks on the beach, and no reason to enjoy the comfort of shade from the soaring heat. Our collective humanity very much depends on the natural world, for joy, for comfort, and for sheer survival. Nature is full of complex and dynamic systems that are constantly interacting with our societies. The natural world is not simply active in some abstract sense; its collective physical interactions guide and forge many fundamental features of human societies and civilizations. Humanity does not exist on a magical pedestal above the rest of reality; we are just one slice in a grand continuum of physical systems that interact, combine, and transform over time. We too belong to the natural world, and we too experience its interactions and conditions, just like everything else. The wonders of the world are waltzing to the rhythm of restless atoms and oscillating fields.

The fact that we belong to nature means that understanding nature is the key to understanding ourselves, and the historical journey of humanity more broadly. One of the major curiosities about the development of our species is why it took so long for human beings to establish energy-intensive civilization. About 50,000 years ago, our ancestors had roughly the same anatomical features as we do today. Their skulls and bodies were similar to ours. They almost certainly had similar intellectual capacities; some human groups were probably speaking fully developed languages.[12] But for all these important similarities, they lived in a very different world than ours. They didn't have agriculture, roads, writing, and large buildings. They didn't have government bureaucracies. Why did they lack these things, even though they had the intellectual capacities to create them? To answer this question, it's useful to know something about their natural environment. For the past two million years, the climate of planet Earth has been dominated by large temperature variations that have produced everything from warm and flourishing conditions to alarmingly frigid patterns in which massive glaciers engulfed much of Europe, Asia, and the Western Hemisphere. This climatic chaos came to an end roughly 15,000 years ago, with the conclusion of the last Ice Age. What followed was a period of remarkable and unusual climatic stability in which the world experienced not just warmer temperatures, but *stable* temperatures, stable rainfall patterns, and stable volcanic activity for prolonged stretches of time.[13] Even this world had its share of sudden variations and cataclysmic natural disasters, of course, but at least people didn't have to deal with most of Europe being buried under massive stacks of ice.

Humans almost certainly knew about the basic patterns of agriculture long before the establishment of fully developed agricultural civilizations. But for tens of thousands of years, that knowledge didn't really matter because we couldn't do much with it. The transition to agriculture was messy, complex, and unfolded over thousands of years; most hunters and gatherers actually resisted such a massive societal change at different points in time.[14] But shifting ecological dynamics provided the possible operating space for agriculture to develop in the first place. The Ice Age had deprived people of the resources and ecological conditions to generate the necessary food surpluses for a sedentary community, which is *not* to claim that sedentary communities are what people wanted. Many of them were perfectly happy being nomadic. But even if they had wanted something else, nature was not exactly cooperative. In a topsy-turvy world of radical climatic shifts, counting on a sedentary location for a reliable food supply was just wishful thinking. A promising settlement in one year could be obliterated the next year by a powerful blizzard or some other natural disaster. People needed to have something far more regular and reliable to survive, much less to establish a successful agricultural community. Before the dawn of agriculture and the domestication of animals, the chaotic climate forced people to rely on food sources that were often seasonal and sporadic. As a result, widespread climatic instability probably delayed the emergence of energy-intensive civilization by tens of thousands of years, even though humans had all the physical and intellectual capacities required for its realization. By contrast, the climatic stability of the last 10,000 years has served as the central ecological primer that allowed human beings to form complex societies and civilizations. A predictable climate allowed people to grow food surpluses from the same settlement. It allowed them to organize their lives and communities without the fear that everything would collapse in a few years. In short, it afforded them the first real opportunity to move away from their previous nomadic existence, although this process came with its own set of problems and complications.

Nature evidently imposes powerful limits on what societies can do. Our civilizations are complex biophysical systems that interact with the wider natural world, and none can be fully examined apart from their underlying material conditions. The main goal of this book is to develop a better theoretical framework for thinking about the links and relations between the natural world and human economies, in the hope that such knowledge can help us plan for a better future and wrestle with the complex problems of the intensifying ecological crisis in our own age. A secondary goal is to offer a specific and alternative worldview for how to construct the broader energetic, economic, and political parameters of future societies. The dynamic structure of economics rests on physical and social conditions. Economies are complex dynamic systems in which people exchange energy with their natural ecozones and interact with one another as they produce, consume, and distribute goods and services. Economic systems can expand or contract depending on underlying ecological conditions, class structures, labor relations, social conflicts, and technological adaptations. These material factors are all locked in a vast web of chaotic interdependence, meaning that changes in one factor can produce highly complex and non-linear effects on the other factors. Ecological conditions may retain their causal primacy in the long run, but social and economic factors can predominate over critical periods that constrain the distribution of financial wealth as well as the intensity of commercial and productive cycles.

I will call this basic summary the *ecodynamic synthesis*. It's not meant to be a formal scientific theory with rigorously quantifiable predictions. The larger aim is instead to present a new worldview that is informed by various intellectual traditions, including scientific and economic sources. The ecodynamic synthesis borrows and combines insights from ecological science, network theory, complex systems theory, and various economic theories, chiefly Marxist and institutional theories. Underpinning the ecodynamic synthesis is a flow-cycle model involving nature and society. Nature is the flowing river and society is the rotating watermill. Dynamical flows are the collections of energy and materials that economies absorb from the natural world. Some portions of these flows are used for production and distribution, but others are lost as wasted energy that can lead to natural instabilities that are highly disruptive to the prevailing social order. Dynamical cycles, on the other hand, are the recurring collective modes of production and distribution facilitated by the flows. The flows constrain and facilitate the cycles by providing a resource base for energy extraction, but the cycles can also impact ecological flows through various chaotic feedback loops associated with the energy losses of our economic activities, and other non-linear perturbations.

Before getting into more details, I should add a few notes about the philosophical orientation of the book. First, it would be wrong to see the message of this work as anything approaching *environmental determinism*, a controversial term that's difficult to define. It roughly means something like the belief that the economic, cultural, and political structures of human society are mostly, if not exclusively, caused by environmental factors and conditions, including the prevailing climate and the available resources in a particular place. Such a reductionist framework will inevitably leave out a lot of important details. The climatic conditions of the United States can never tell you why there are specifically 50 states in the country, or 435 members in the House of Representatives. The river systems and arable lands of the United States cannot explain why we need to file our tax returns with the Internal Revenue Service by April of every year, nor can they explain the specific tax rates, credits, and deductions that applied to us in those returns. There are many legal, institutional, and economic features of society that are simply *emergent*, that cannot be reduced to deeper scientific principles through naïve models of causation. However, it's still true that the global ecosphere *sets the broader constraints in which human societies and civilizations can successfully operate*. And if these dynamical constraints experience profound and rapid disruptions, as they are in our age, then the natural world can indeed compel and pressure human civilization into reorganizing its emergent properties. Likewise, the emergent features of human society can also be helpful or disruptive to the rest of the natural world, depending on what we do. That's the basic point of this book: nature and society can affect each other in highly complex ways, which means that we need to think carefully about what kind of society we want to have in the first place.

To speak about "nature" and "society" might seem to imply that these two terms signify fundamentally different things. But isn't society just a part of nature? Society is embedded in the natural world, as are all human bodies, feelings, thoughts, and ideas. In that case, why would we talk about them dialectically, as if they're two separate things? The answer is simple. We need to distinguish between a *system* and its *surroundings* if we're going to say anything empirically and scientifically useful about the system itself. Even though a car engine forms one part of a larger vehicle, and that vehicle is embedded in a particular environment, it's still useful to distinguish between the engine as a system and the wider environment in which the engine needs to operate. Only then can we start making claims about the efficiency or the work output of the engine, among other things. If we want to understand the long-term evolution of human society, then we can only do so by exploring how society interacts with the rest of the natural world. Much of this exploration will be done from the perspective of physics, but it's important to emphasize that this book is not just about physics. It's also about the social, economic, political, and historical relations that inform the wider scientific discussions regarding humanity's future on this planet. The goal is to aim for a holistic analysis that ties together different causal factors. Physics will be the centerpiece of this analysis, but the book will also relate these scientific ideas to other emergent features of human society.

Energy

The fate of all economic systems is written in their energy flows. The exchange of energy between different systems has a decisive influence on the order, phase, and stability of physical matter. Energy is

INTRODUCTION

typically defined as the ability to do work, but this definition is not ideal when considering things like radiation or quantum physics, so one can better think of it as any conserved state of motion, such as work or heat, that can be exchanged among different systems.[15] Kinetic energy and potential energy are two of the most important forms of energy storage. The sum of these two quantities is known as mechanical energy.[16] A truck speeding down the highway packs a good amount of kinetic energy, which is energy associated with motion. A boulder teetering at the edge of a cliff has great potential energy, or energy associated with position. If given a slight push, its potential energy transforms into kinetic energy under the influence of gravity, and off it goes. When physical systems interact, energy is converted into many different forms, but its total quantity always remains constant. The conservation of energy implies that the total output of all energy flows and transformations must equal the total input.

Energy flows among different systems represent the engine of the cosmos, and they happen everywhere, so often that we hardly notice them. Heat naturally flows from warmer to colder regions, hence our coffee cools in the morning. Particles move from high-pressure areas to low-pressure areas, and so the wind starts to howl. Water travels from regions of high potential energy to regions of low potential energy, making rivers flow. Electric charges journey from regions of high voltage to regions of low voltage, and thus currents are unleashed through conductors. The flow of energy through physical systems is one of the most common features of nature. As these examples show, energy flows require gradients, which are differences in temperature, pressure, density, or other factors. Without these gradients, nature would never deliver any net flows, all physical systems would remain in equilibrium, and the world would be very boring. Energy flows are also important because they can generate mechanical work, which is any macroscopic displacement in response to a force.[17] Lifting a weight and kicking a ball are both examples of performing mechanical work on another system.

Energy flows are central to the existence of life. For example, living organisms rely on the conversion of chemical energy into thermal and mechanical energy for survival. Living things usually store chemical energy in large biomolecules called adenosine triphosphates. Every single day, life is hard at work producing vast quantities of ATP molecules from proton pumps powered by electrochemical gradients across cell membranes, a fancy way of saying that energy flows in our cells produce the things we need to survive. Proteins and enzymes then break down these biomolecules and exploit the resulting energy to perform vital biological tasks, such as the contraction of muscles or the replication of DNA. Another example is the radiation zooming out from the Sun, which is generally responsible for some of the most important energy flows in the circle of life. A long time ago, the ancestors of many special organisms learned how to harness the power of sunlight and other basic substances for survival. Today these organisms include plants and algae; biologists call them *autotrophs*.

In the energy pyramid of life, the autotrophs are the base because they sustain all other living things. Part of the energy they process and absorb makes its way to *heterotrophs*, the organisms that need the nutrients provided by autotrophs to function and survive. Herbivores, carnivores, and omnivores are all heterotrophs, ultimately reliant on plants and other autotrophs for their basic needs. Herbivores consume plants directly while carnivores evolved to consume herbivores as a way of acquiring the substances they need from plants. Nuclear fusion inside the Sun is what keeps the whole system going. A byproduct of nuclear fusion is radiation, packets of photons that wend their way through the Sun before being released into space. These photons carry tiny amounts of energy individually, but vast quantities collectively. As they strike the Earth, they excite electrons and other particles to higher energy states, setting off a series of reactions and energy flows that help to sustain the chemical and biological features of life. The Sun is the chief sustainer of all the relevant energy flows in the biosphere.

Entropy and Dissipation

Although energy flows can produce work, they rarely do so efficiently. Large macroscopic systems, like trucks or planets, lose or gain mechanical energy through their interactions with the external world. The lead actor in this grand drama is dissipation, defined as any process that partially reduces or entirely eliminates the available mechanical energy of a physical system, converting it into heat or other products.[18] As they interact with the external environment, physical systems often lose mechanical energy over time through friction, diffusion, turbulence, vibrations, collisions, and other similar dissipative effects, all of which prevent any energy source from being converted entirely into mechanical work. The presence of dissipation makes it necessary for dynamical systems to regularly absorb energy from the external world, as a way of overcoming the energy losses associated with dissipative interactions. Car engines burn through gasoline to sustain the motion of the wheels, which need periodic bursts of energy to overcome the dissipative effects of friction. Another simple example of dissipation is the heat produced when we rapidly rub our hands together.

Dissipation is not limited to things like cars and machines. It can have stunning and profound effects on planetary and cosmic scales. For example, the friction between the tides and the surface of the Earth is gradually slowing down the rate of rotation of the entire planet. Around 600 million years ago, a day on Earth lasted about 22 hours, and 300 million years later it had gone to almost 23 hours. As the tides move across the crust, the resulting friction produces heat. Over the eons, the ultimate effect of this process has been to slow down the rotation of the planet, which loses rotational kinetic energy as it produces more heat. In the natural world, macroscopic energy flows are often accompanied by dissipative losses of one kind or another. Physical systems that can dissipate energy are capable of rich and complex interactions, making dissipation a central feature of the natural order. A world without dissipation, and without the interactions that make it possible, is difficult to imagine. Think about what would happen if friction suddenly disappeared from the world. People would be slipping and sliding everywhere. Our cars would be useless, as would the very idea of transportation, because wheels and other mechanical devices would lack any traction with the ground and other surfaces. You would never be able to hold hands with your loved ones. You could never rock your baby back and forth. Our bodies would rapidly deteriorate and lose their internal structure. There would be no basic concept of time and place, not in a traditional sense anyway. The world would be very alien and unrecognizable.

Dissipation is closely related to *entropy*, one of the most important concepts in thermodynamics. While energy measures the motion produced by physical systems, entropy tracks the way that energy is distributed in the natural world. Entropy has several standard definitions in physics, all of them essentially equivalent. One popular definition from classical thermodynamics states that entropy is the amount of heat energy per unit of temperature that becomes unavailable for mechanical work during a thermodynamic process.[19] Another important definition comes from statistical physics, which looks at how the microscopic parts of nature can join to produce big, macroscopic results. In this statistical version, entropy is a measure of the various ways that the microscopic states of a larger system can be rearranged without changing that system.[20] For a concrete example, think of a typical gas and a typical solid at equilibrium. Energy is distributed very differently in these two phases of matter. The gas has a higher entropy than the solid, because the former's particles have far more possible energy configurations than the fixed atomic sites in solids and crystals, which have only a small range of energy configurations that will preserve their fundamental order.[21] We should emphasize that the concept of entropy does not apply to a specific configuration of macroscopic matter, but rather applies as a constraint on the number of possible configurations that a macroscopic system can have at equilibrium.

Entropy has a profound connection to dissipation through one of the most important laws of thermodynamics, which states that heat flows can never be fully converted into work.[22] Dissipative interactions ensure that physical systems always lose some energy as heat in any natural thermodynamic process, where friction and other similar effects are present. Real-world examples of these thermodynamic losses include emissions from car engines, electric currents encountering resistance, and interacting fluid layers experiencing viscosity. In thermodynamics, these phenomena are often considered irreversible. The continuous production of heat energy from irreversible phenomena gradually depletes the stock of mechanical energy that physical systems can exploit. According to the definition of entropy, depleting useful mechanical energy generally implies that entropy increases. Formally stated, the most important consequence of any irreversible process is to increase the combined entropy of a physical system and its surroundings. For an isolated system, entropy continues to rise until it reaches some maximum value, at which point the system settles into equilibrium. To clarify this last concept, imagine a red gas and a blue gas separated by a partition inside a sealed container. Removing the partition allows the two gases to mix together. The result would be a gas that looks purple, and that equilibrium configuration would represent the state of maximum entropy. We can also relate dissipation to the concept of entropy in statistical physics. The proliferation of heat energy through physical systems changes the motion of their molecules into something more random and dispersed, increasing the number of microstates that can represent the macroscopic properties of the system. In a broad sense, entropy can be seen as the tendency of nature to reconfigure energy states into distributions that dissipate mechanical energy.

The traditional description of entropy given above applies in the regime of equilibrium thermodynamics. But in the real world, physical systems rarely exist at fixed temperatures, in perfect states of equilibrium, or in total isolation from the rest of the universe. The field of non-equilibrium thermodynamics examines the properties of thermodynamic systems that operate sufficiently far from equilibrium, such as living organisms or exploding bombs. Non-equilibrium systems are the lifeblood of the universe; they make the world dynamic and unpredictable. Modern thermodynamics remains a work in progress, but it has been used successfully to study a broad spectrum of phenomena, including heat flows, interacting quantum gases, dissipative structures, and even the global climate.[23] There is no universally accepted meaning of entropy in non-equilibrium conditions, but physicists have offered several proposals.[24] All of them include time when analyzing thermodynamic interactions, allowing us to determine not just whether entropy goes up or down, but also how quickly or slowly physical systems can change on their path to equilibrium. The principles of modern thermodynamics are therefore essential in helping us understand the behavior of real-world systems, including life itself.

All life forms are engaged in the physical process of avoiding thermodynamic equilibrium with the rest of their environment by continuously dissipating energy. This was one of the key insights that the physicist Erwin Schrödinger suggested in the 1940s, when he used non-equilibrium thermodynamics to study the central features of biology.[25] We may call this vital process the *entropic imperative*. All living organisms consume energy from an external environment, use it to fuel vital biochemical processes and interactions, and then dissipate most of the energy consumed back to the environment. The dissipation of energy to an external environment allows organisms to conserve the order and stability of their own biochemical systems. The essential functions of life critically depend on this entropic stability, including functions like digestion, respiration, cell division, and protein synthesis. What makes life unique as a physical system is the sheer variety of dissipation methods that it has developed, including the production of heat, the emission of gases, and the expulsion of waste. This sweeping capacity to dissipate energy is what helps life to sustain the entropic imperative. Indeed, the physicist Jeremy England has argued that physical systems in a heat bath flooded with large amounts of energy can tend to dissipate more energy.[26] This "dissipation-driven adaptation" can lead to the spontaneous emergence of order, replication, and self-assembly among microscopic units of matter, providing a potential clue to the very dynamics of the origin of life. Organisms also use the energy they consume to perform mechanical work by, for example, walking, running, climbing, or typing on a keyboard. Those organisms with access to many energy sources can do more work and dissipate more energy, satisfying the central conditions of life.

Like all other biological organisms, humans consume resources from some external environment in order to survive, reproduce, and expand. The consumption of energy is crucial because it allows animals to perform mechanical work, which means that they can run away from predators, move to other areas in search of food, or build tools to solve different problems. In addition, the biological mechanisms that are central to life, like protein synthesis and cell division, all require energy conversions in order to unfold at a regular pace. And the energy that we consume from the external world is precisely what allows our cellular components to perform all of these critical tasks. Energy and work are measured in the same units, the standard one being the *joule*. One joule is roughly the amount of energy your hand uses to lift an ordinary apple up to your mouth. A *food calorie*, also known as a *kilocalorie*, is equal to about 4,184 joules. A central empirical focus of our future analysis will be the rate of energy use, mostly because we want to track how the exchange of energy between different systems can change over time, and how those changes then affect the structural organization of the systems themselves. The rate of energy use is known as *power*, but to avoid any confusion with the more familiar understanding of the word, we shall use terms like "power output" or "rate of energy conversion" whenever we talk about power. In addition, we want to remove population effects in our analysis by using the term "energy conversion" to mean *per capita* conversion, unless otherwise indicated.

Physics and Economics

The thermodynamic relationships among energy, entropy, and dissipation impose powerful constraints on the behavior and evolution of economic systems.[27] From a thermodynamic perspective, economies function as non-equilibrium systems capable of rapidly dissipating energy to some external environment. All dynamical systems gain strength from some energy reservoir, reach peak intensity by absorbing a regular supply of energy, then unravel from internal and external changes that either disrupt vital energy flows or make it impossible to keep dissipating more energy. They can even experience long-term undulations by growing for some time, then shrinking, then growing again, before finally collapsing. Interactions between dynamical systems can produce highly chaotic results, but energy expansions and contractions are the core features of all dynamical systems. The energy consumed by all economic systems is either converted into electricity and mechanical work, along with the physical products derived from that work, or it's simply wasted and dissipated to the environment. Economies that increase the amount of electricity and mechanical work they generate can usually produce more goods and services. Historically, electricity and mechanical work have comprised a relatively small fraction of total energy use in any economy; the vast majority of the energy consumed by all economies is routinely squandered to the environment as waste, dissipation, and other kinds of energy losses.

Throughout history, economic growth has depended heavily on people converting more energy from their natural environments and concentrating the resulting energy flows for the completion of specific tasks.[28] When humans were hunters and foragers, the primary asset that performed mechanical work was the human muscle.[29] Our muscles were great for running, walking, collecting fruits, and hunting animals, but they did not generate enough mechanical work for energy-intensive production. This nomadic way of life lasted for some 200,000 years before undergoing significant disruptions after the Ice Age. Over the next few millennia, new lakes and rivers were born from the receding glaciers. Flora and fauna rapidly multiplied; this bonanza of wild crops and animals made it possible for some groups of people to settle down on specific plots of land, leading to numerous pastoralist and agricultural strategies that would come to define much of our history. Agrarian economies relied heavily on cultivated plants and domesticated animals to help generate surpluses of food and other goods and resources. These agrarian modes of production and consumption dominated human societies for almost ten thousand years but were eventually replaced by a new economic system: capitalism.

Capitalism emerged and spread from a complex set of converging conditions: the extraction and development of energy-dense natural resources, waves of colonial expansion, intensive periods of industrialization and technological innovation, the proliferation of epidemic diseases, and genocidal campaigns against Indigenous populations accompanied by massive land theft. It was the most chaotic and violent transformation in human history. The global economy has since become an interconnected system of finance, computers, factories, vehicles, machines, wage labor, and much more. Creating and sustaining this system required a major upward transition in the rate of energy throughput from our natural environments. In our nomadic days, the daily rate of per capita energy consumption was around 5,000 kilocalories, perhaps even less.[30] By 1850, per capita consumption in Britain had risen to roughly 80,000 kilocalories per day, and the rate has since ballooned to about 200,000 kilocalories a day among the most advanced energy-intensive economies, such as the United States.[31]

From a scientific perspective, the fundamental feature of all capitalist economies is a high rate of energy use focused on boosting productivity and economic growth. The collective deployment of machines, vehicles, and electronic devices requires the production of vast amounts of useful energy, which in turn allows people to make more stuff, travel farther distances, and lift heavier objects, among other tasks. Capitalism is far more energy-intensive than any previous economic system, and it has wrought unprecedented ecological consequences that may threaten its very existence. We can think of capitalism, from a biophysical perspective, as a supercharged entroplex, a mega-dissipative system dumping massive amounts of gases, liquids, and solid waste into our natural environments. The biggest effect of this spasmic energetic release has been the degradation of the planetary ecosphere to a more entropic and chaotic state, with profound implications for the future of humanity. In 2004, a group of research scientists with the International Geosphere-Biosphere Programme summarized these enormous changes and the great acceleration that started in the middle of the twentieth century:

> A profound transformation of Earth's environment is now apparent, owing not to the great forces of nature or to extraterrestrial sources but to the numbers and activities of people—the phenomenon of global change. Begun centuries ago, this transformation has undergone a profound acceleration during the second half of the 20th century. During the last 100 years human population soared from little more than one to six billion and economic activity increased nearly 10-fold between 1950 and 2000. The world's population is more tightly connected than ever before via globalization of economies and information flows. Half of Earth's land surface has been domesticated for direct human use. Most of the world's fisheries are fully or overexploited. The composition of the atmosphere—greenhouse gases, reactive gases, aerosol particles—is now significantly different than it was a century ago. The Earth is now in the midst of its sixth great extinction event. The evidence that these changes are affecting the basic functioning of the Earth System, particularly the climate, grows stronger every year. The magnitude and rates of human-driven changes to the global environment are in many cases unprecedented for at least the last half-million years.[32]

INTRODUCTION

The increasing order of human civilization is producing increasing disorder in the broader ecosphere, and this growing imbalance, if allowed to continue, will inevitably cripple the stability of global civilization itself. The huge energy losses of our modern economies have become an energy reservoir for other dynamical systems in nature, such as viruses, bacteria, hurricanes, storms, wildfires, and algae blooms, among others.[33] Paradoxically, the "useless" energy that human civilization dumps into the natural world powers the formation of other physical systems, and these systems collectively are forming a new ecological order that will be incompatible with the necessary conditions for sustainable human development. It remains uncertain how long we can sustain such an energy-intensive path, but there is no doubt that the fantasy of endless growth and easy profits cannot continue. Our main goal is to chart a new path that will allow humanity to thrive within the natural constraints and parameters of the planetary ecosphere.

Our Great Challenge

In the 1980s, the world had a major ecological problem. The emission of gases like chlorofluorocarbons (CFCs), widely used as coolants in refrigeration, had punched a giant hole in the ozone layer. If that hole continued to get bigger, it would mean that dangerous levels of ultraviolet radiation from the Sun would reach the surface of the Earth and wreak havoc on human health worldwide. The world's nations got together and passed the Montreal Protocol in 1987, which gradually phased out the use of CFCs in refrigeration and other commercial or industrial uses. Even though it still hasn't fully gone away, the hole in the ozone layer gradually became smaller over the next few decades. Overall, this episode is often cited as prime evidence of humanity's collective ability to solve hard problems. Here were the countries of the world all coming together in the face of adversity and doing something for the common good. What's often neglected in this feel-good story is that CFCs were then immediately replaced by substances like hydrofluorocarbons (HFCs), which are extremely potent greenhouse gases. Humanity essentially traded one problem for another: it got rid of CFCs and saved the ozone layer, but only at the cost of dumping vast amounts of HFCs into the atmosphere and causing a rapid acceleration in global warming.

The most dangerous recurring theme of human history is civilizational collapse and dysfunction caused by widespread ecological disruption. We are now entering an age of profound and unprecedented ecological crisis, a *bionomic disruption* that threatens not just the viability of our economic systems, but large portions of the planet's biological fabric. Although global warming is a major long-term challenge, it's just one of many fundamental problems that have been caused by the capitalist acceleration of the past two centuries. For just one notable example, recent studies have suggested that roughly 8 to 10 million people die prematurely every year from air pollution caused specifically by the combustion of fossil fuels.[34] In addition to global warming, humanity has adversely affected the planetary ecosphere by polluting the world's oceans and atmosphere, starting the sixth mass extinction event in our planet's history, causing massive disruptions to the nitrogen and phosphorous cycles that sustain all plant growth, boosting the incidence of global pandemics through agricultural expansion, and expediting the shortage of critical natural resources, including drinking water, for much of the world's population. To the extent that human civilization will buckle and bend over the next few centuries, it will be from the cumulative pressure of all these factors, not from any single factor by itself. But since our problems are largely framed in the context of climate change, the corresponding solutions are narrow and technical, amounting to little more than hoping for rapid technological change to replace fossil fuels with renewables like wind and solar. For this school of thought, the goal is simply to decarbonize the global economy and continue as if nothing else matters. This approach is misguided precisely because our ecological problems don't boil down to one simple thing. Substituting fossil fuels with renewable energy sources is

INTRODUCTION

an important and admirable objective, but not if it's done without thinking about the broader context of our ecological challenges.

The dominant narrative in this entire debate is the idea that humanity can overcome the ecological problems of late-stage capitalism through technological innovation.[35] In this book I will make a very different argument. Technological innovation has produced more energy-intensive societies with destructive ecological consequences. Improving energy efficiency tends to lead to more energy use over time, not less. Having a good understanding of the physics of capitalism will provide a better foundation for radically changing the politics of capitalism. Instead of advocating for technological change, we should advocate for social and political transformation. We should change how power is wielded in society, how decisions are made, how labor is organized, how wealth is distributed. Revolutionary transformation is what we need to secure our future. The main thesis of this work can be stated as follows: if modern civilization is going to survive the upcoming millennium, then we need to thoroughly reconstitute and revolutionize the social and economic relations that currently govern our lives. There are no quick technological "fixes" for what we're facing; technological changes can supplement a comprehensive solution to our bionomic dilemma, but the only realistic way to save civilization in the long run is to radically change how we organize our societies.

In the first four chapters, I will lay out some of the concepts, problems, and narratives that have been used to understand the interconnections between human economies and the natural world. In chapters 5 through 9, I'll unveil a new theoretical framework for better understanding economics and our place in the natural world. In the last three chapters, I'll use this framework to identify some of the possibilities that are worth pursuing in our future efforts to construct a genuinely post-capitalist world.

Among all the living organisms that have called this blue marble home, humans are a very recent species. In that short period of time, we have managed to become one of the most dominant life forms in the history of the planet, creating powerful civilizations with elaborate cultures, large populations, and extensive trade networks. We have been nomads and farmers, scientists and lawyers, nurses and doctors, welders and blacksmiths. Our achievements are both astonishing and unprecedented, but they also carry great risks. The economic and demographic growth of human civilization over the last ten thousand years has profoundly impacted natural ecosystems throughout the planet. Global civilization now stands at a critical tipping point that deserves closer scrutiny. The goal of this work is to explore the deep ecology of human life and existence: to develop a fundamental sense of how human society relates to the wider natural world and then to thoroughly examine the possible scenarios for the future of human civilization. If we are to have any hope of addressing the difficult challenges we face, we must begin by understanding them and appreciating their complexity. And then, we must act.

CHAPTER 1

Growth and Scale in Economics

It's a truism that our world is based on economic growth. Politicians and economists constantly talk about it. "Economic growth is good for our country because of the jobs it creates and the prosperity it spreads," according to former U.S. Treasury secretary John W. Snow.[1] Another former Treasury secretary, Steve Mnuchin, claimed that "our most important priority is sustained economic growth and I think we can absolutely get to sustained 3% to 4% GDP, and that is absolutely critical for the country."[2] Political scientist Michael Mandelbaum echoes the same basic idea: "Economic growth is necessary to keep the promise . . . that each generation will have the opportunity to become more prosperous than the preceding one."[3] There's near-unanimity in elite circles that growth is not just a good thing, but, more broadly, that it should be the central organizing principle of economics and of human society.

But what does it mean for an economy to grow? How should we understand the concept of economic growth? These questions are not just some silly academic exercises. If we're going to figure out how to make our economies compatible with the wider planetary biosphere, then we need to impose constraints and specify parameters for acceptable levels of growth and scale. And if we're going to talk about growth and scale, then we need to have a good sense of what's actually growing and how big it's getting, which means we need to address this question: How should we measure the size of an economy? This deceptively simple question is actually quite tough to answer. In an economic context, size and scale can mean lots of different things. For example, we can measure the size of a country's population and see how it changes over time. We can measure the total mass or volume of everything produced by a particular society. We can measure a society's total energy consumption. What we should measure and call the "size of an economy" really depends on what we're trying to study. If we're looking at crop productivity in agriculture, for example, we might measure the caloric content produced per unit area. The specific demands of the problem should guide the parameters of the definition.

In the reigning elite discourse, typified by Mnuchin's comment above, there's apparently only one measure of economic scale: *Gross Domestic Product* (GDP). It's hard to find another number that's used more often to provide ideological justification for capitalist expansion. For economists and politicians who support capitalism, "economic growth" has come to mean a positive change in GDP. But while many people in our capitalist world are obsessed with the idea of growth, there's very little reflection about whether GDP is the proper way of measuring it. In reality, GDP epitomizes and exacerbates all of the tensions, contradictions, and intellectual acrobatics that dominate contemporary economic discourse. The fact that severe conceptual and mathematical issues exist with using GDP as an infallible empirical metric seems to bother very few mainstream economists.

In macroeconomics, the academic field that studies the supply and demand cycles of entire economies, GDP has become the quasi-absolute gold standard, the rubric that supersedes and seemingly explains all other economic metrics. GDP can be calculated through several methods that, in principle, should yield the same answer, although it's not an easy number to get. Government agencies work long and hard trying to measure and estimate the spending patterns in various economic sectors, then they add up these sectoral results to get the total value for the entire economy. And then there's the issue of interpretation, the issue of how to understand what GDP means exactly. Many economists and organizations will tell you that GDP measures *aggregate demand*, the total amount of money exchanged for final goods and services; that is, we count the final sale price of a car but not the price of the metal used in the car's production. In this framework, GDP supposedly represents something like an index for the market value of final production. It serves as a quantity for measuring the productive "real output" of society. The Organization for Economic Cooperation and Development flatly defines GDP as "an aggregate measure of production equal to the sum of the gross values added of all residential institutional units engaged in production."[4] The International Monetary Fund states that "GDP measures the monetary value of final goods and services…produced in a country in a given period of time. It counts all the output generated within the borders of a country"[5] In his popular college textbook, *Principles of Economics*, the economist Gregory Mankiw defines GDP as "the market value of all final goods and services produced within a country in a given period of time."[6]

Evidently, then, GDP is supposed to be a monetary indicator of society's aggregate productive output. That's the primary reason why most economists use it and cite it so frequently: because it apparently reveals something fundamental about a society's total production. Economists also generally believe that GDP correlates very well with certain biophysical factors, such as life expectancy, infant mortality, or energy consumption. They point to these correlations as evidence that GDP is picking up something meaningful about what's happening in the physical realm of production, in the real world beyond money. Opponents of the supremacy wielded by GDP in our public debates typically respond in several ways. First, they point out that GDP calculations ignore the consequences of economic activity and treat all spending equally. If a hurricane comes by and destroys a city, the money required to rebuild the city is added to the GDP. Ecological degradation is ignored, in other words, and even treated as a net plus when we spend money to recover from the degradation that we caused in the first place. One could be forgiven for thinking that economists have arbitrarily settled on a metric that deliberately ignores the downsides of growth and consumption. If these downsides happen to provide large drags on future economic performance, then counting the costs associated with them as positive is downright ridiculous.

Second, GDP is blind to many important social and economic activities that are not financially rewarded. For example, it leaves out the value of domestic housework and domestic childcare, even though both are critical to a well-functioning society. Women worldwide perform critical domestic labor that is not financially rewarded or valued by their respective societies. Third, GDP says nothing about how economic growth is distributed. This point is especially important for the recent history of capitalism, in which the dominant nations of the global economy have entered prolonged periods of stagnation where the rich continue to hoard more of the income generated by production—thus increasing their wealth—at the expense of almost everyone else. And contrary to what most economists believe, GDP per capita does not often correlate well with quality-of-life factors. Cuba now has a comparable life expectancy to that of the United States, but Cuba's GDP per capita is about ten times smaller. This is not just an isolated case; it's a very common feature in the data for many countries, as we'll shortly see.

The measurement of GDP also suffers from innumerable boundary problems, which are related to decisions about what kinds of economic activities to include and exclude in GDP figures. In 1987, for example, Italy decided to incorporate much of its shadow economy into its GDP. That decision meant including many informal and illegal economic activities in the national accounts. Italian GDP instantly increased by 18 percent, and Italy suddenly had a larger economy than Britain, an event the Italians dubbed *il sorpasso*, or "the overtaking."[7] In 2014, Britain itself included sales from illegal drugs and prostitution in its GDP totals, part of a broader trend among European nations to include more illegal economic activities in their national accounts.[8] But not all countries included the same things; France, for example, did not add proceeds from prostitution services to its total GDP.[9] In the 1990s, the System of National Accounts maintained by the United Nations, the IMF, and the World Trade Organization started allowing certain kinds of income made by banks to be included in GDP figures. Over the years, more and more types of financial services have been added to GDP numbers. In principle, GDP numbers are supposed to be measuring spending on goods and services, but as history clearly shows, not everyone agrees about what the word "services" actually means. For example, stock purchases are generally omitted from GDP figures, but aren't the investors buying up these stocks providing an important service to the corporations that are getting the money?

Even if economists recognize the limits of GDP, they certainly seem unwilling to share this message with the rest of society. The capitalist press, itching to make huge profits by blasting out empty headlines under tight deadlines, has detached GDP from any kind of rational economic analysis and bastardized it into the universal measure of social and economic progress. When the economist Simon Kuznets developed the preliminary methodology for the calculation of GDP, he cautioned that it should never be used as a metric for social welfare or social progress. "The welfare of a nation," Kuznets warned, "can scarcely be inferred from a measurement of national income."[10] The very person who invented the ideas behind GDP understood the severe limitations of national income figures. But his warning has gone largely unheeded. In the capitalist press and in official government publications, GDP figures are often treated like divine gospel. GDP has become a standard marker for just about any discussion related to big social questions. Things have gotten bad enough that if we take GDP and divide it by population, we have apparently found a society's *standard of living*, according to the elite circles vested in the survival of capitalism.[11] It doesn't end there. Take total energy consumption and divide it by GDP and we apparently have a society's *energy intensity*.[12] Take the inverse of this last quantity and we suddenly have a country's *economic efficiency*.[13] Take greenhouse gas emissions and divide them by GDP and we've amazingly discovered a nation's *emission intensity*.[14] Just about everywhere one looks, GDP serves as both an excuse and a reference frame for some silly calculation that's meant to reveal something profound about the world. The goal in this chapter is to scrutinize GDP more closely and to explain some of its basic flaws, and then to offer alternative ways of thinking about the aggregate scale and growth dynamics of modern economies.

Units and Prices

The fundamental reason why GDP cannot accurately estimate "real" economic output is because *prices are not stable units of measurement*. Simply put, changes in GDP do not tell us *anything* about economic growth or decline. To really unpack this claim, we need to know something about how *units* relate to *measurements*. Suppose that a friend measures your height one day by using a tape measure marked in feet and inches. You find out that you're 6 feet tall, as expected. Exactly a week later, you tell your friend to measure your height again, just for fun. But this time your friend decides to trick you. Instead of using a normal tape measure, he picks one with a different value for the inch. In this modified tape measure, an inch has been redefined to be *twice as big* as the standard inch, which is about 2.54 centimeters. The new inch is therefore 5.08 centimeters. What do you think that your friend will measure now? Because the new inch extends over twice as much distance as the old inch, it will take *half* as many "new inches" to cover your entire height. Your friend will tell you that you're only 3 feet tall; your height has gone down dramatically in just a week! Of course, we all know that's not really true. Nothing about your physical height actually changed. The new height reported by your friend simply differs from the old height because your friend changed the unit of measurement the second time around. This little story highlights an important lesson: for good and reliable measurements of physical quantities in the real world, the units used to do the measuring *must be stable over time and space*.

In the core sciences, like physics and chemistry, units are extremely important. Physicists go to great lengths to accurately define their units.[15] The meter, for example, is officially defined as the distance covered by light in a minuscule fraction of a second: 1 divided by roughly 299.79 million.[16] This underlying stability in the definition of the meter is important precisely because it allows for equivalent comparisons. With a stable meter, it makes sense to say that a tower measured in meters is taller than a horse measured in meters. As long as the meter means the same thing for the horse as it does for the tower, then comparisons between the two in meters remain valid. Given that the definition of a meter is tied to a fundamental constant of nature, the speed of light, we can be very certain that the tower will remain taller than the horse at all times, if we are measuring both in meters. But imagine that we lived in an alternate reality where the value of the meter was always shifting for different objects. If the meter ever had one definition for the horse and another for the tower, then trouble could quickly ensue. Suppose we start off with an equivalent meter for both and measure the height of the tower at 50 meters and the height of the horse at 2 meters. But next year, the distance measured by the tower meter doubles while the meter for the horse falls by a factor of 50. Then we would measure *the horse* at 100 meters and the tower at 25 meters. The horse has become "taller" than the tower. These examples highlight the importance of stability for our units of measurement.

A stable unit allows us to make sensible comparisons by

guaranteeing that it can be added up in constant amounts, or "aggregated." If we measure any system with an unstable unit, then we are introducing the possibility that the measured values of the system could be changing *because the unit itself is changing*, not because the system actually became larger or smaller. Aggregation is fairly straightforward in the natural sciences because the underlying units of analysis are very stable and well-defined. In a toy example, suppose we have a box of 20 red marbles and 10 blue marbles, with each red marble having a mass of 100 grams and each blue marble having a mass of 500 grams. Finding the total mass of our box is easy: simply record the mass of the empty box, add up the masses of all the red marbles, add up the masses of the blue marbles, and then add these numbers together to get the total mass. If five years pass by and someone tells us that the system has the same number of marbles and the same distribution between red and blue marbles, then the total mass would still be the same. If someone came along and added 30 identical red marbles to the system, we would simply add the mass of these 30 red marbles to the total mass calculated five years ago. If the red marbles were all replaced by blue marbles, aggregating the new total is still easy because the mass ratio of a blue marble to a red marble is constant. The reliability of aggregation in this simple example is guaranteed by the fact that the definition of a gram is stable and constant over time. It provides a fixed weight for aggregation, which implies that any change to the total mass of this system must be due to whether the total number of red marbles and blue marbles changed or that the distribution of red marbles and blue marbles changed. In other words, changes in the aggregate mass of the box reflect actual physical changes in the number or distribution of the constituent marbles that make up the system. In a topsy-turvy world where people kept changing the definition of a gram for each of the two marble types, aggregation would be useless because then changes in the total mass of the system would depend not just on the number of marbles in the box but also on changes in the definition of a gram as well. In other words, changes in the measured mass of the box would not be a pure reflection of changes in the physical world; they would also reflect *shifting social conventions that have been inserted into the measurement process.*

In the world of economics, things work very differently compared to the natural sciences. Real-world economies contain many different kinds of commodities, from cars and shoes to books and sweaters, that cost money to purchase. And when it comes time to add things up, it's precisely the money that economists are actually measuring and aggregating, not the numerical counts of different goods and services. Let's consider how this affects GDP at a very high level. Economists use a variety of methods to calculate GDP, such as the *income approach*, which simply adds up the income flowing to all economic agents, or the *expenditure approach*, which adds up the spending done by those same economic agents. In the expenditure approach, for example, GDP is calculated by summing up government spending, consumer spending, the trade balance, and business investment. These methods all have something in common: they're adding up sales, and sales depend on prices. What's the problem with that? The problem is that prices are *not* stable units of measurement because they change in unpredictable ways over time, which means that they cannot provide a fixed weight for aggregation. With certain "adjustments" that supposedly remove the distortions caused by changing prices, economists seriously pretend that they can aggregate using money in the same way that physicists aggregate using meters and kilograms. This is the absurd fabrication to which we now turn: the idea that monetary units can objectively measure "quantities of stuff" as a kind of proxy for something in the real world, like levels of production.

Imagine that you wanted to measure the annual production of an apple farm. An easy way to do it is by measuring the total weight of the apples harvested in a given year. And if you wanted to compare the productivity of one apple farm to another, you would just see which farm harvested the higher total weight, adjusting the result for farm size and other factors. But you'll run into a big problem if you try the same strategy for other kinds of companies and economic activities in the real world. We'll call it the *aggregation problem*, and it's the problem of how to add up a bunch of things that are very different from one another by *using a common unit of measurement*.[17] In the real world, there are many different types of commodities and services, which means that we somehow need to figure out a common unit of measurement *for all of them*, assuming that we want to aggregate and find a total value for production in the entire economy. Suppose you wanted to compare the annual productivity of your apple farm to the annual productivity of a law firm. Using weight as the standard of measurement now seems ridiculous because it's very difficult to know what you should be *weighing* for the law firm. Apple farms produce a specific output: apples. By contrast, law firms provide a wide array of legal and consultation services that cannot be easily quantified in natural units. Economists believe that they have a solution: find the total value of these different goods and services by using prices as a universal benchmark. In other words, track the annual sales of the apple farm and add them to the annual sales of the law firm. If you wanted to compare the productivity of the two companies, then just look at which one has the higher sales, adjusting for workforce size and other factors along the way. At a superficial level, prices appear to solve the aggregation problem. In reality, however, prices are unstable units of measurement and are absolutely incapable of providing a reliable method of aggregation. *Any aggregate quantity measured in prices can change either because the quantity of stuff being measured is changing or because the prices themselves are changing.* Aggregating and comparing using prices is like measuring the horse and the tower with a changing meter stick.

Most economists believe that price fluctuations both reveal and obscure the "real" output of the production process. They reveal it in the sense that price changes supposedly reflect supply and demand dynamics in the real economy. But they also obscure it in the sense that they only capture productive output indirectly, hence the need to adjust financial aggregates for the effects of price fluctuations. Ideally, economists would love to know the monetary value of what the economy produced without worrying about these fluctuations. That's why generations of economists have been more than happy to flatten and obscure the effects of heterogeneity and to ignore the chaotic and differential nature of market prices. In 1956, for just one example, the economist Robert Solow declared in a famous paper describing his theory of economic growth that "there is only one commodity, output as a whole."[18] He made this obviously false and ridiculous assumption so that he could "speak unambiguously of the community's real income." In the real world, of course, there isn't just one commodity, there are trillions of them with radically different price vectors and trajectories, so the notion of a "real income" is pure fantasy.

Economists usually calculate the "real" value of output by adjusting for *inflation*, the general rise in prices for goods and services across the economy over time. Inflation typically occurs in response to the convergence of various social, political, technological, and ecological factors, such as wars, pandemics, government fiscal and monetary policies, and corporate pricing strategies. These are some of the common causes behind high inflation, but of course there are many others, up to and including rampant speculation and price gouging during unstable economic times by those who control a particular market. The causal dynamics of inflation and deflation, along with their associated impacts on the biophysical scale and the ecological consequences of an economy, will be explored in further detail in the second part of the book. For now, the focus remains on whether price changes are an accurate indicator of economic growth and scale, not necessarily on what's causing them in the first place.

Adjusting for inflation seems necessary in order to prevent the monetary value of production and consumption from artificially changing because of fluctuating prices, which do not always indicate that greater quantities of goods and services are actually available. For example, if a farmer produces and sells 5 apples this year for $10 each, she'll have $50 of sales. But she might be able to produce and sell 5 apples next year for $15 each, reaching a higher sales figure of $75 even though she didn't produce any additional apples. This is where economists would say, "Adjust the new sales figure for inflation" by holding the prices constant. Although such an adjustment would work for this simple example, it turns out that it does not work *in general*, as we're about to see with some concrete examples. That's because in the real world there are trillions of different prices floating around, often with multiple prices for the same product. A gallon of gas might cost $3.50 in one gas station while another one down the street might sell it for $3.80. The same toy might cost $20 at Target and $10 if you buy it online with Amazon. In the real world, prices are highly variational and differential, and so are *changes in prices across time*. Because it's impractical to track trillions of different prices and adjust for each individually, economists basically create fancy weighted averages of sampled goods and services and call that a *price index*. These price indexes are supposed to measure some kind of imaginary price level in the economy, and adjusting a particular financial aggregate value by that price level is supposed to give us the "real output" of the economy; in other words, it's supposed to give us the financial value of production after adjusting for inflation. This adjusted financial value is meant to reflect actual changes in production or other changes in the real world; it's not supposed to be clouded or distorted by changing prices. That's the hope anyway.

The reality is different. The reality is that prices are unpredictable and chaotic, and relative price changes between commodities do not follow stable patterns. This fact is extremely important: market prices are not just unreliable units of aggregation simply because they change over time. The really crucial point is that these changes are *chaotic and non-uniform*. If all market prices bounced up and down in unison, then we'd have no aggregation problem at all. But because market prices change over time in *unpredictable ways*, they cannot provide a fixed basis for aggregation. Because this feature makes it *impossible* for prices to serve as a fixed unit of aggregation in the real world, economists have typically responded by holding prices constant from a particular time in the past. That way they can just *pretend* that prices are a stable unit of measurement, but at the cost of picking arbitrary weights for the resulting aggregation. We may call this issue the *problem of time*, which is the problem of deciding what specific price weights from the past should be used in any process of financial aggregation. Different price weights will yield different answers for the aggregate values.

Consider the average monthly price changes for bananas and Red Delicious apples in the United States, as shown in Figure 1.1 The figure shows data from January 2003 to January 2010 compiled by the Bureau of Labor Statistics for the Consumer Price Index (CPI). It's easy to see from the chart that apple and banana prices are not changing in tandem. For an egregious example, look at what happens in the 45th month after January 2003: apple prices skyrocketed by nearly 30 percent while banana prices fell by 1 percent. Once you dig through the data and find the prices you want, you decide to add up the apple and banana sales to figure out the "total output" of this imaginary two-commodity economy. You follow the path of the economist and perform this addition while holding prices constant, to "remove" the distortions caused by fluctuating prices. Of course, the prices are *not actually constant*; they're changing every month by different percentages. But you're going to *pretend* they're constant by picking a particular point in time and aggregating *only* with prices from that point. The problem should now be immediately apparent. Depending on which apple and banana prices are chosen, you will give a greater relative weight in the aggregation either to the apples or to the bananas. If you happen to pick that crazy month where apple prices go up 30 percent, then "total productivity," as measured in monetary terms, will be more influenced by the price change among the apples rather than the price change for the bananas. And if you pick prices from another month, you'll get an entirely different answer for productivity. The lesson of this story is simple: *Because there are many ways to adjust for inflation, there can never be a single monetary value of production.*

FIGURE 1.1: Average monthly percent changes in apple and banana prices, January 2003 to January 2010

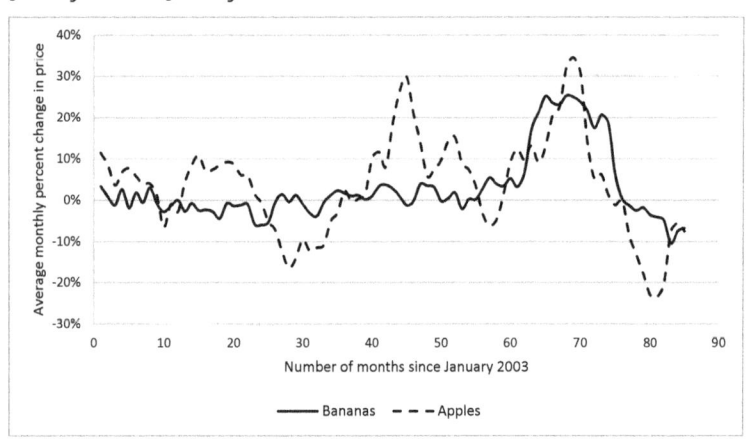

Source: Consumer Price Index data, Bureau of Labor Statistics, https://www.bls.gov/cpi/.

Figure 1.1 is emblematic of how relative prices change through chaotic patterns over the years. The price of an apple relative to the price of a banana in any given year usually differs compared to the same ratio calculated in previous or in future years. Because commodity prices are unstable and chaotic, they do not provide a fixed and constant weight for measurement. Prices depend on time, and what they reveal about economic activity or production also depends on the time that we pick to evaluate the prices. Picking different times would yield different price weights, and different price weights would tell us different things about what's happening in the economy. Aggregate quantities constructed through unstable units, like prices, *do not have a single, well-defined value at every instant in time.* They can have multiple values for the same time, and those values depend on which price weights are used for the aggregation. Something so profoundly ambiguous and ill-defined cannot serve as a reliable standard for empirical measurement.

The aggregation problem is absolutely fatal for the core argument that most economists want to believe about money, which is that monetary dynamics reflect something fundamentally objective about the "real" state of the world beyond social relations. But the aggregation problem is not the only issue for inflationary adjustments. Like all other averages and aggregates, inflation indexes have their own boundary problems. For example, the media often obsesses over the changes in the Consumer Price Index (CPI), which measures price changes for most consumer goods and services in the United States but does *not* say anything about changes in asset prices, like stocks, bonds, or real estate property. It's a convenient form of propaganda for the rich, because while asset inflation skyrocketed between the Great Recession and the onset of the Covid pandemic, hardly anyone in the corporate media saw it as a crisis. It's not hard to understand why: the vast majority of assets are owned by the rich, so when asset prices rise dramatically, the rich are much better off. But the converse is that when trillions of dollars are shoveled toward jacking up asset prices, the rest of the economy suffers from underinvestment, low wages for common people, and depressed levels of demand. Consider that, for the 449 publicly listed companies in the S&P 500, about 54 percent of their earnings from 2003 to 2012 went into stock buybacks, a staggering sum of roughly $2.4 trillion.[19] Dividend payments to shareholders ate up another 37 percent of their earnings. That left only 9 percent of profits for things like investments or boosting wages for employees. To take a more concrete example, the top ten semiconductor companies in the United States spent a total of $168 billion on purchasing their own stocks from 2010 to 2021.[20] Intel alone accounted for $86 billion in that period. From 1990 to 2021, Intel bought a massive $152 billion of its own stocks, an annual average of roughly $5 billion. Imagine what Intel could have done if they'd been more strategic with the way they invested just some of that money; maybe they wouldn't be the corporate beggar they are today, waiting for the federal government to give them a handout at every turn. Excessive levels of asset inflation, often called *asset bubbles*, are in many ways more dangerous than excessive levels of consumer inflation, because the former is rarely accompanied by a substantial increase in nominal wages and makes it extremely difficult for most people to climb up the economic ladder. In other words, asset bubbles under modern capitalism tend to fix and ossify class relations, reinforcing the status quo.

Basic Examples

Let's consider a few simple numerical examples to understand why adjusting nominal monetary aggregates for inflation does not work, in general. Before looking at a case that doesn't work, we're going to look at one that does: an idealized world with a single commodity. Suppose you produced 5 apples in a given year and sold each of them for $10. You made a grand total of $50 in sales. In the next year, you increase the price of an apple to $20, a price increase of 100 percent, and sell 6 apples. Now you made $120 in sales. Of course, this kind of jarring price increase is unrealistic, but these numbers were chosen to keep the calculations simple and straightforward. The same conclusions would follow if you picked smaller price changes. Next, you adjust your production for inflation by taking that $120 and discounting its value back to the first year. The discount rate is 100 percent, the same as the price increase of the apples. When you reduce a number at a rate of 100 percent over a year, you're effectively reducing its value by half. Why is that? Think about the opposite scenario. What happens when you increase a number by 100 percent? You're basically *doubling* its value. If you add *any* number to itself, the answer you get is twice the original number. Add 5 to 5 and you get 10, which is 5 times 2. Add 20 to 20 and you get 40, which is 20 times 2. In our case, every apple is going from costing $10 in the first year to costing $20 in the second year. To *undo* the effects of this doubling in the second year, we have to reduce our new sales figures by half. In other words, half of $120 is $60, which is the inflation-adjusted value of our productivity. Notice that $60 divided by $50 from the first year is the *exact same ratio* as the 6 apples produced in the second year divided by the 5 apples. In other words, our inflation-adjusted monetary value changed by the same fractional amount as the number of apples we produced. Great news!

GROWTH AND SCALE IN ECONOMICS 49

Now let's consider a slightly more realistic scenario: a world with apples and bananas. Imagine that everything about the apples stays the same as in our first example. For the bananas, suppose that we produced 5 in the first year and sold them for $10. In the second year, we produce 10 of them and sell them for $8 each. In the first year, we sold $50 worth of apples and $50 worth of bananas, for a grand total of $100. In the second year, we sold $120 worth of apples and $80 worth of bananas, for a grand total of $200. Now we have to adjust for inflation again, but the problem is that we have to somehow combine two different price changes, one for the apples and one for the bananas, into a single *aggregate* index. Why can't we just adjust for inflation separately, splitting the apples and the bananas into different components? We certainly can, but at the cost of having two different measures of inflation. Adopting a separate inflation index for every single commodity in existence would leave us with billions, and potentially *trillions*, of different inflation rates. We can kiss goodbye to the idea of a "price level" for the entire economy! This strategy would definitely worry most economists, who make their living on the belief that it makes sense to talk about such an aggregate "price level."

On this basis, then, we are forced to create an aggregate index for inflation. But how do we do that? The most direct method is to calculate the average price change. Other methods also exist, and we're going to encounter one of them shortly, but let's keep things simple for now. In our example, apple prices rose by 100 percent, from $10 to $20, and banana prices declined by 20 percent, from $10 to $8. The average of 100 percent and negative 20 percent is 40 percent. With our inflation rate on hand, we adjust the $200 in the second year to get the "real" value. Notice that when you increase the value of a number by 40 percent in a single year, it's equivalent to multiplying the original number by a factor of 1.4. To undo the effects of this increase, we have to divide our $200 of sales by 1.4, which gives us an inflation-adjusted total of $143. When we divide $143 of sales in the second year by $100 of sales from the first year, we get a ratio of 1.43. Does this answer match the ratio of total fruits produced in the second year to total fruits produced in the first year? The answer is no.

In the second year, we produced 16 fruits, 10 bananas and 6 apples, while in the first year we made 10 fruits, 5 bananas and 5 apples. The ratio of 16 divided by 10 is 1.6, meaning that the "real" monetary value increased by a *lower* rate than the rate of change for what we actually produced. Notice that if we had produced the exact same number of bananas as apples in both years, and sold all of them at the exact same price in both years, then everything would work out. In other words, when commodity prices and quantities change *uniformly*, it makes sense to adjust for inflation. But in the real world, commodity prices and quantities *diverge* from each other in highly chaotic ways, as we've already seen. In our example, notice that the price change for the bananas was different than the price change for the apples, and the number of bananas we produced in the second year was different than the number of apples we produced in the same year. This is exactly what happens in the real world: relative prices and quantities diverge in very unpredictable ways, as we saw with the Bureau of Labor Statistics data in Figure 1.1. This basic fact makes it impossible to create a consistent inflation index, forcing us to make extremely arbitrary choices about how to measure inflation in the first place.

Suppose we adopt a different method to adjust our "production" for inflation, one that's actually popular among economists: holding the prices from a particular base year constant. A base year is just some past year when the prices of these two commodities are known. The idea is to calculate the total monetary value of our commodities by using prices from a single year, even if we produced the commodities in other years as well. Let's run through the examples above with this new method to see if we have any luck. In our single-commodity example with the apples, holding the first-year prices constant gives us $50 of sales in the first year (5 times $10) and $60 of sales in the second year (6 times $10). Because we have two years of sales, we have two possible choices for the base year. Suppose that we now hold the second-year prices constant. Our total sales would become $100 in the first year (5 times $20) and $120 in the second year (6 times $20). The good news is that the ratio of $120 to $100 is the same as the ratio of $60 to $50. In both cases, the monetary value increased by 20 percent. But this is purely an artifact of choosing an idealized world with one commodity. Now let's look at the example with apples and bananas. We start off by holding the first-year prices constant. Total sales in the first year are equal to $100 (5 apples times $10 plus 5 bananas times $10). Total sales in the second year come out to $160 (6 apples times $10 plus 10 bananas times $10). When we switch base years, holding second-year prices constant, our sales in the first year come out to $140 (5 apples times $20 plus 5 bananas times $8). Sales in the second year are equal to $200 (6 apples times $20 plus 10 bananas times $8). When we use the first year to hold prices constant, our monetary proxy for production rises by 60 percent, going from $100 to $160. But when we use the second year as our base year, the monetary aggregate goes from $140 in the first year to $200 in the second year, an increase of only 43 percent. The "real" monetary aggregate increases by different amounts depending on which base year we pick to adjust for inflation. Essentially, our indicator of production is telling us two different things about the change in output. It turns out that this method simply reproduces the problem we encountered earlier. The aggregation problem still stands.

One might be tempted to think that the first method in the preceding paragraph provides a definitive solution, since the aggregate financial quantity for that case increased by 60 percent, the same relative increase as the total number of fruits produced. Apparently, all we have to do is use the prices from the first year of the series and we're all done. This path is actually a mirage, however. In the real world, we don't have only two years of sales. We have apple and banana prices going back hundreds of years, and that's just for prices stated in dollars; price data for other historical currencies might go back thousands of years for many commodities. In such a scenario, it's difficult if not impossible to know what to consider as the "first year" of the series. Economists would also not be very fond of adjusting nominal financial aggregates in the present with base-year prices from hundreds of years ago, simply because those historical prices are extremely low relative to current prices. The consequence of this move is that all inflation-adjusted financial aggregates in the present, including GDP figures, would become very tiny—so tiny that they could call into question nearly all of the ideological narratives that support the current economic order. For this and other similar reasons, economists tend to adjust nominal financial quantities with prices from *recent* base years, thereby ensuring a relatively mild adjustment to the nominal figures. But now we're back to the same problem we faced before: how do we choose the base year among multiple competing options that are all valid and sensible? There's no compelling reason to pick one base year over another, which means that we still won't know the "real rate" of change in productivity.

The Varieties of GDP

The aggregation problem is not just a theoretical curiosity. Multiple national governments throughout the world, from India to Nigeria, have artificially changed their "real" GDP figures by simply switching base years. The case of Nigeria is particularly shocking, or amusing, depending on your point of view. In 2014, the West African nation managed to nearly *double* its GDP just by switching base years from 1990 to 2010.[21] Through a simple statistical gimmick, Nigeria suddenly became the largest economy in Africa! Nothing changed on the ground in the lives of actual Nigerians; it's not like overnight they had twice as much stuff as before. It's not like Nigeria was using twice as much energy overnight. It's not like the country's population doubled overnight. There were no comparable biophysical changes in Nigeria to justify that kind of shift to the country's economic scale. The government simply made the social and political decision to change the unit of measurement, thus making it appear as if the size of the country's economy had dramatically increased. Of course, had the Nigerian government made a different choice for the base year, then Nigeria's economy might have doubled, increased by 50 percent, gotten smaller, or anything in between! It's impossible to say because it's all a silly statistical game with massive political implications. We should emphasize again that Nigeria is not alone; these cheap tricks are common throughout the world. There's little doubt that these changes happened for any other reason than desperate politicians eager to quote a bunch of useless numbers to their constituents as a way of showing "progress," that most Orwellian word under capitalism. Fiddling around with base years can produce huge disparities and uncertainties in the corresponding GDP figures. In 2019, the economist Blair Fix used data from the United States government to determine that there's a "30 percent uncertainty in the growth of US GDP per capita over the last 60 years."[22] Unsurprisingly, the official numbers released by the federal government always stayed at the upper end of the estimated range of values.

In the 1990s, the Bureau of Economic Analysis (BEA) in the United States switched its underlying methodology for calculating GDP. Instead of acknowledging the impossibility of objectively adjusting aggregate output for inflation in any real-world economy, it simply adopted a supposedly clever method called "chain-weighting." The basic idea behind chain-weighting is to replace the fixed base year with a moving base year. In the original method, picking an arbitrary and fixed base year could abruptly change "real" GDP even when nothing changed in the production process itself. The BEA believed that using a moving target would minimize the scale of these changes. In its new method, we need to pick a time relative to which the base year can move, and that time is called the "reference year." If the reference year was last year, for example, we would use a formula to calculate a price index for the current year and the previous year. As mentioned before, a price index is meant to be an aggregate measure of the "price level" for a basket of commodities, relative to some point in time. There seems to be no shortage of price indexes among economists; there are literally dozens of them around. The United States government itself uses about a dozen different formulas and methods for a variety of purposes. The BEA actually performs its GDP calculation by combining price indexes and quantity indexes, which are intended to track quantity changes in commodity production instead of changes in prices.

The most popular price and quantity indexes include the Laspeyres index, the Paasche index, and the Fisher index, which takes the geometric mean of the first two. Again, it's important to emphasize that economists have created literally dozens of price and quantity indexes, which is proof in and of itself that there's no one magical method for measuring inflation. Economists usually say that the Laspeyres price index tends to overestimate the inflation rate. Meanwhile, the Paasche index tends to underestimate inflation. The Fisher index is supposed to come in and save the day, slicing right through the middle and minimizing the biases of the other two indexes. An important assumption underlies these claims, the assumption that an aggregate rate of inflation, a "true" rate, actually exists above and beyond the methods that we use to measure it. Price indexes are just supposed to approximate this "true" rate as much as possible. Unfortunately, economists have no way of measuring this profound truth independently of their statistical artifacts. Measuring inflation is not like measuring the speed of light. The latter actually exists independently of human observers. Experimental observations of the speed of light might dance around the real value a little bit, but at least *we know that there is a real value*. By contrast, the rate of inflation is a hazy construct that depends on highly subjective choices about what constitutes a price change for an entire basket of goods. Trying to measure a "true" rate of inflation forces us to confront the inescapable aggregation problem, which means that we have to figure out how to properly weight price changes and how to account for quality variations between different commodities (for example, if a product's quality deteriorates, this is treated in a sense the same as an increase in price). We lack any concrete solutions to these problems.

There is no "true" rate for the Fisher index to approximate. The Fisher index itself is a statistical fiction, a lazy way to harmonize two biased indicators through a cheap mathematical trick. These biases also appear in chain-weighted GDP figures. Economists begin the chain-weighting process by calculating the Fisher quantity index in each year, going all the way back to the reference year. The indexes are then all multiplied together to determine the total index, which is used to adjust the nominal dollar value of GDP for the current year. If the reference year was three years ago, for example, we would calculate a quantity index for the current year and the previous year, then another quantity index for the previous year and the year before that, and finally a quantity index for two years ago and the reference year. We would then multiply the three indexes to get the total index. This total index would serve as the measure of how much output and production have changed over the last three years.

Although chain-weighting might be an improvement over the earlier base-year method, it still ignores the fundamental "problem behind the problem," which is the futility of using unstable units of measurement for aggregation. We still need to pick a reference year for chain-weighting to work, and the reference year that we pick will affect the calculated value of the GDP. The reference year determines the number of Fisher indexes that have to be "chained" together; an earlier reference year means more indexes are multiplied together whereas a later year means fewer indexes are multiplied together. The basic issue is still there: we have an ill-defined aggregate being measured through an unstable unit. And there is no sound empirical or theoretical reason to pick one reference year over another. All of them are equally valid, which means that the vast ambiguities surrounding GDP figures cannot be removed.

Economists like to emphasize that the chain-weighting method preserves GDP growth rates even when the reference years change. The idea is that switching to a new reference year may change the actual value of "real" GDP, but at least it won't affect the growth rate series, thus making economic history seem more stable. In the older base-year approach, changing the base year meant changing both "real" GDP and the growth rate of "real" GDP over time. It sounds like a good defense, but it doesn't work at all. First, this argument only holds assuming that we're just using one type of index. But if economists switched from the Fisher index to the Walsh index, for example, then the growth rates of the GDP time series would most definitely change. Another problem with the argument is that it's only valid if the chaining period remains fixed. The BEA currently chains together *quarterly* Fisher indexes, but there's no fundamental reason why it can't chain indexes with monthly, semiannual, or other kinds of temporal weights. When you change the temporal weight, the GDP growth rates also change. Contrary to the standard dogma, chain-weighting does not necessarily preserve growth rates. By using a rolling average as a glossy finish, chain-weighting has simply become another method of hiding and obscuring the aggregation problem.

To concretely see why chain-weighting resolves nothing of importance, let's look at the databases on chain-weighted GDP maintained by the Federal Reserve. The Federal Reserve keeps a vintage database of chain-weighted GDP using 2009 as a reference year, as well as the next iteration of the figures with a reference year of 2012. Figure 1.2 shows a comparison of "real" GDP calculated from these two reference years from 1947 to 2016, the latest year available with the 2009 benchmark. Let's pick the year 2014 as an example. What was the "real" GDP of the United States back then? If you adjust nominal GDP in 2014 with a reference year of 2009, then the "real" GDP of the United States in 2014 was about $16 trillion. But if you do the adjustment with a reference year of 2012, then the "real" GDP of the United States in 2014 was $16.9 trillion. That's a difference of almost $1 trillion for reference markers that are only three years apart! This gaping discrepancy has

GROWTH AND SCALE IN ECONOMICS 57

FIGURE 1.2: Chained Real GDP of the United States with two different reference years, 2009 and 2012

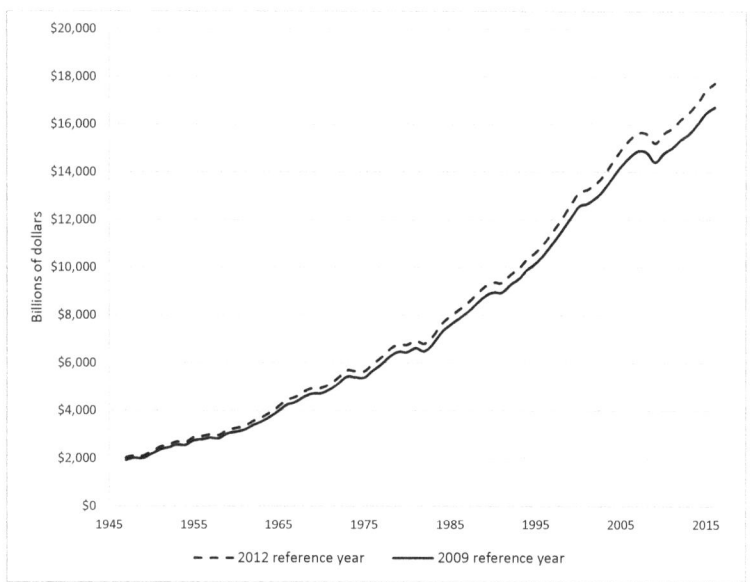

Note: The GDP figures shown here are annual averages.
Source: Federal Reserve Economic Data. For the discontinued 2009 series, see https://fred.stlouisfed.org/series/GDPMC1#0. For the 2012 series, see https://fred.stlouisfed.org/series/GDPC1#0.

far more significance than just being intellectually interesting. It brings down most of the fantasies surrounding GDP and economic development. Once there's no longer a single "real" GDP number for a single year, then there's also no singular GDP per capita for that year, and gone is the notion that we can determine a country's standard of living based on GDP. As Figure 1.2 makes absolutely clear, this problem is not limited to the year 2014. It's endemic throughout the entire series, and it's particularly bad for years in the twenty-first century.

The central point of this analysis is that there is no such thing as the "real" GDP of the United States in any given year. Depending on the method of aggregation, *any particular country can have thousands of real GDP numbers in any given year*. The notion that there's a single, fixed monetary aggregate that tells us something profound about the scale or the productivity of an economy is pure fantasy. In that case, why do governments only report a single number for "real" GDP in their official publications? Because the government economists doing these calculations simply pick an arbitrary method of aggregation and call it a day. In the United States, for example, the BEA and the Federal Reserve update the reference year in periodic cycles. They're currently adjusting nominal GDP figures with a reference year of 2017, then in the future that marker will move to later years.[23] There is no grand or meaningful argument to justify any of these particular choices. The government could easily adjust nominal GDP with a reference year of 2013 or 2004 if it wanted to. Even though the resulting numbers are completely arbitrary, they're presented as if they indicate something important about the production levels of the national economy.

It's worth noting that mainstream economists do sometimes acknowledge that real GDP cannot actually measure "real output," whatever that is. But they might respond to these critiques by claiming that GDP provides a "relative" index of the market value of production. In other words, GDP tells us something important about the production of an economy relative to a specific point in time. But why should anyone care about that? Just substitute the word "arbitrary" for the word "relative" and you'll get the same result. Measurements of real GDP are *necessarily arbitrary*. Here is the economist Charles Steindel in a 1995 paper about the measurement of GDP:

> The economy consists of millions of individuals and firms producing a multitude of goods and services. This complexity virtually ensures that any method of estimating "real GDP" involves making some more or less arbitrary decisions about the most appropriate way to add up data from individual sectors.[24]

The BEA also acknowledges the aggregation problem in chapter 4 of its *NIPA Handbook*, writing the following:

The fundamental problem confronting the efforts to adjust GDP and other aggregates for inflation is that there is not a single inflation number but rather a wide spectrum of goods and services with prices that are changing relative to one another over time. The index numbers for the individual components can be combined statistically to form an aggregate index, but the method of aggregation that is used affects the movements of the resulting index.[25]

In this instance, both Steindel and the BEA are absolutely correct: real GDP is an arbitrary quantity. It does not and cannot tell us anything meaningful about the physical process of production.

As if these issues aren't bad enough, using dollars to specify the GDP of any other country besides that of the United States also brings up the problem of exchange rates. If you adjust a country's nominal GDP in its national currency and then convert that figure into dollars, the conversion will depend on the exchange rate between the dollar and the other currency. After Russia's invasion of Crimea in 2014, for example, the Russian ruble fell sharply in value against the dollar because of Western sanctions, and Russia's GDP measured *in dollars* also collapsed. But again, not much changed in the lives of ordinary Russians. The country experienced a mild economic downturn after the imposition of Western sanctions and then easily recovered. The exchange rate problem alone is bad enough, because it prevents us from doing straightforward comparisons among national GDP figures, even in the absence of the aggregation problem.

Some economists have tried to overcome this problem by focusing on *purchasing power*, an approach that has been increasingly adopted by organizations like the United Nations.[26] The basic idea behind the method is that a dollar can get you different quantities of goods and services in different countries. If a typical haircut in the United States costs $15, a typical haircut in India might only set you back a dollar. The dollar has greater purchasing power in India than in the United States. Economists pick different "baskets" of goods and services and then count how many of these items the dollar can buy in different countries. They use this data to construct purchasing power ratios between two economies, and they adjust GDP figures based on these ratios. Unsurprisingly, the method is riddled with problems, such as deciding which basket of goods should be used for comparison and the fact that these international comparisons don't account for any quality differences in the products being considered. All we're left with is a steaming pile of confusion. In many media reports, for example, you'll read that the United States is the biggest economy in the world. But in many others, you'll find that somehow China is actually the biggest. The discrepancy arises because different media organizations pick different standards. If you go with the standard exchange rate method, the United States is the biggest. If you adjust for purchasing power, then it's China. Does the exchange rate method for calculating GDP correspond to biophysical realities on the ground in China and the United States? Let's see. In the early 2020s, China produced almost 60 percent of the world's steel, about 75 percent of the world's electric vehicle batteries, and roughly 80 percent of the world's solar panels.[27] In 2021, the United States consumed roughly 93 exajoules of energy compared to China's sky-high 158 exajoules, which is roughly 70 percent more than the U.S. figure.[28] By any sensible readings of aggregate scale and production, it's obvious that the Chinese economy is indeed far bigger than the American one. But GDP comparisons between the two countries, even those based on purchasing power, do not accurately reflect these gaping differences in biophysical quantities.

Although real GDP is a useless quantity from the perspective of production, *nominal* GDP does have several helpful properties that justify its continued calculation: it clarifies the compositional structure of spending in the economy and it provides critical details about the investment landscape under capitalism. To better understand the social and financial dynamics of the capitalist system, it's nice to know how much money businesses are investing in retail or in agriculture in any given year. It's nice to know how much money consumers are spending on movies versus restaurants. It's nice to know the size of government spending relative to private sources of spending in the economy. It's nice to know these things because they shed light on the dynamic social relations of the capitalist system. Sectors and industries that are doing well, that are dominating the market in various ways, will probably employ a greater share of the population and will likely have more political power. For these and other reasons, the compositional information that we gain from measuring financial quantities as a share of GDP is quite valuable, and I will indeed make many references throughout this book to various quantities expressed as a fraction of GDP. But we don't need to pretend that these numbers tell us anything fundamental about the objective productive output of the economy.

As a last-gasp effort to save their most precious metric, some economists will quickly claim that GDP per capita is highly correlated with important quality-of-life factors. Even if it's a flawed metric, the argument goes, it's still capturing something important about the state of our societies and economies. Life expectancy and infant mortality rates are two of the most popular examples brought up in these lines of argument. I'll focus on the former because the same points apply for the latter as well. The relationship between GDP per capita and life expectancy is often called the *Preston curve*, after the demographer Samuel Preston. I've reproduced it in Figure 1.3 for many countries around the world in 2015, and one can see that it shows an extended arc where life expectancy generally rises with higher GDP per capita levels.

There are two important takeaways from this chart that tarnish the supposedly strong association between life expectancy and GDP per capita. First, going past a certain GDP per capita doesn't seem to produce further improvements in life expectancy. Second, and far more important, there are absurdly large variations in life expectancy outcomes at the lower ranges of GDP per capita. This result should definitely not be the case in a world where the GDP metric serves as a powerful causal driver of improvements in life

FIGURE 1.3: GDP per capita vs. life expectancy for countries in 2015.

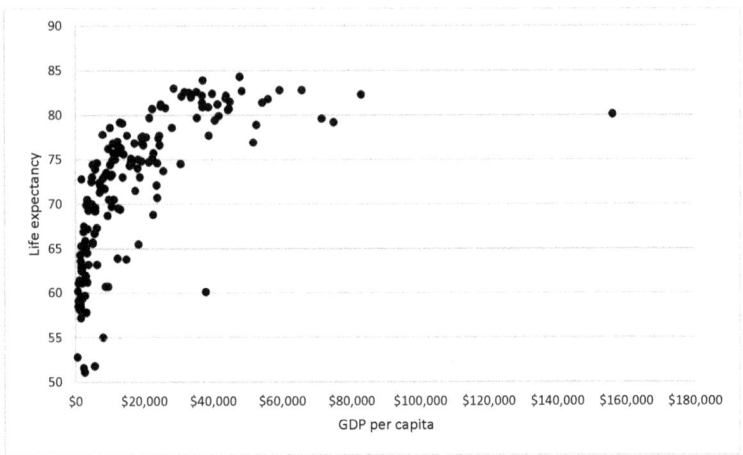

Source: The data shown here comes from the public data source Our World in Data, available at https://ourworldindata.org/life-expectancy.

expectancy. One can make many glaring comparisons to highlight this absurdity. Consider Honduras and Nigeria. They had a very similar GDP per capita in 2015, roughly $5,000, and yet life expectancy in Honduras was 73 years while in Nigeria it was only 52 years. Evidently, GDP per capita cannot be the causal explanation for this huge difference because it's effectively the same for both countries. For another notorious case, consider Cuba and the United States. Cuba had a very similar life expectancy to the United States even though its GDP per capita was just 15 percent of the American one. China had a life expectancy of 77, two years below the American figure, but with one-quarter of America's GDP per capita. It's worth noting that, as of 2022, China actually had achieved a higher life expectancy than the United States for the first time in history, a consequence of the Covid pandemic and other social and political failures in the United States.[29] One could go on and on here, but the central point is clear: GDP per capita is neither a significant causal driver nor a compelling explanation for life expectancy. Better explanations for these results can be found by focusing on how much nations invest in education, healthcare, and infrastructure, regardless of their GDP per capita numbers. For example, Cuba has historically had high life expectancy figures because the Cuban government invests quite heavily in the country's healthcare system.

How Scale and Efficiency Affect Growth and Collapse

Conventional economic theories tell us precious little about aggregate production and its growth, largely because they have no reliable way of measuring either one, but also because many of them do not acknowledge the existence of system dynamics in the real world. Any non-equilibrium thermodynamic system that goes through phases of growth must also go through phases of decline and collapse. What goes up must come down. On the empirical side of things, the issues presented above collectively underscore the inescapable dangers of using financial aggregates as straightforward barometers of economic productivity. Money can never tell us who makes more stuff or who has the biggest economy, because money is not measuring something objective in the real domain. Money is a symbolic facade that human beings created to organize their social world for certain ends. It's an emergent property of complex societies, and the monetary quantification of commodities reveals far more about the class and political dynamics behind the social process of *distribution* than it does about the physical process of *production*. Distribution is the proper context for analyzing the nominal domain, which is the world of social relations quantified through monetary indicators. In other words, the nominal domain is the realm of things like prices, incomes, profits, wages, debts, assets, and interest rates.

In our times, capitalists manipulate the nominal domain in order to control and to organize society as they see fit. They set wages and prices to levels designed to boost profits and to beat out competitors. Of course, the *social* aspects of the production process are profoundly interconnected with the results of economic distribution, as Marx brilliantly recognized in Volume I of *Capital*, where he explained how capitalists rigged and perverted the production process to boost profits. Capitalists routinely attempt to suppress wages, to extend the length of the working day, and to use automation as a way of replacing or controlling labor with additional machines.[30] By reducing the financial costs of production at the expense of their workers, capitalists get to walk away with more money for themselves and their investors. But here the specific focus is on the *physical* nature of production, the only aspect of production that can be said to represent "real output," which must be measured and understood *independently* of financial valuations. If they're actually interested in studying this kind of production, economists should stop wasting time with aggregate production functions constructed through monetary units, like GDP, and should start exploring more scientific quantities that are actually worth measuring, such as energy flows and biophysical measures of social progress. Energy is, indeed, the currency of the entire universe, and knowing how it relates to our societies should be the first goal of economics.

Understanding the limitations behind inflation-adjusted GDP values has further implications for ecological studies, especially ecological economics. There's no shortage of research studies in this field that aim to use various thermodynamic quantities to "explain" real GDP, or other financial aggregates.[31] The correlations between these quantities supposedly reveal something profound about the economy's productivity. In reality, they do no such thing. Given how they're constructed from unstable units, monetary aggregates simply *cannot tell us anything definitive about production*. The best way to empirically measure and understand an economy's productivity is by analyzing its energy flows, efficiencies, and qualities. Only these metrics can accurately tell us what an economy is *actually making and doing*, as opposed to the prices and wages that the dominant classes have *ritually and artificially imposed* on the economy for the purpose of controlling whatever it's making and doing.

Energy is not the last word on productivity. We also need to stop pretending that productivity, or any other concept in economics, has some kind of fixed, eternal, and Platonic meaning that exists above and beyond the empirical problems we encounter in the world. As I clarified at the start of this chapter, productivity is a concept that can be understood in different ways depending on the situation that one is trying to analyze. Certain notions of productivity might be more useful or practical than others. One conception of productivity might explain a certain fact about the world better than another, but the latter could still be useful in other contexts. For example, a tentative definition of production that combines the social and physical dimensions of macroeconomics could be the following: productivity is the conversion of certain energy forms to other energy forms that are considered useful in various social and ecological conditions. In this biophysical sense, economic growth can then be seen as *the expansion and diversification of useful energy forms*. In other words, economic growth is *an increase in the amount of energy converted to useful forms by a society in a given year*, along with *the diversified application of useful energy for the completion of different tasks*. This definition emphasizes that productivity is not just about "using more energy to make more stuff," but also concerns the different ways in which we use energy to achieve different social goals. Production is a complex process embedded within human society, and thus its organization is affected by numerous political factors, social conventions, and class conflicts.

Beyond the empirical problems of how to measure and define things like growth and productivity, conventional theories also have issues in recognizing the complex dynamics of real-world economic and social systems. Societies don't always keep growing by producing more stuff. Sometimes they go through unpredictable phases of growth and decline. And sometimes they collapse altogether. It's precisely in this arena where we need new conceptual directions for understanding how economic systems actually work. A great place to begin is by examining the struggle between efficiency and scale, a struggle that is one of the decisive mediators for economic growth and decline. Efficiency gains and technological innovations tend to favor the creation of larger energy systems. That's because in its early phase of expansion an economic system converts energy to produce more stuff: roads, bridges, buildings, aqueducts, factories, and power plants. As it gradually becomes more efficient at making these things, the energy costs per unit required for the system's upkeep and maintenance decrease, meaning that the system has more useful energy available for expansion. Greater efficiency under capitalism is typically used to generate more growth, meaning that it's used to scale up and expand systems of production and distribution, such as building larger factories and facilitating greater volume flows in the global trading system. As the ecological scientist Carey King put it: "In networks, the operational energy consumption, structure, and size are inherently linked. *The more efficient a system's distribution network, the larger it can be.* In short, size matters."[32] In the current era of energy-intensive and late-stage capitalism, economic decision-making is concentrated in the hands of a few elites looking to boost financial profits and to dominate global markets. Scaling up production through efficiency gains is one of the most common ways of doing that in the history of capitalism, since it lowers unit production costs and allows corporations to gain a competitive advantage in global markets by selling their products and commodities at relatively cheaper prices.

This pattern of efficiency-driven growth cannot continue forever, however. That's because as the system becomes bigger, energy use is increasingly diverted toward maintenance, repairs, and upkeep. The more stuff we make, the more stuff we have to maintain over time, and it takes energy to maintain that stuff. In effect, greater scale also makes it harder for a system to keep growing at a rapid rate because the system needs to devote more energy and resources toward upkeep and maintenance on what has already been built or developed, which leaves less energy in total available for expansion. This argument is true even when efficiency and technological improvements are considered. Although efficiency gains help to facilitate expansion, the sheer scale of a large system provides a powerful barrier to faster growth, or sometimes to any growth at all. Bigger and more complex systems need to devote an increasing share of their efficiency gains toward maintaining their complexity, leaving fewer resources available for further rapid growth. Paradoxically, increasing complexity is both a gateway to the initial expansion of a system's energetic scale and a powerful barrier to any further expansion in the future.

In his 1988 book *The Collapse of Complex Societies*, the anthropologist Joseph Tainter examined the downfall of the Maya civilization and the collapse of the Roman Empire. He argued that as societies become more complex, they may become better at solving certain problems, but new ones will come up that pose insurmountable challenges. For example, the rapid growth of the Roman Empire meant that new problems developed related to the management of conquered territories. Because the Romans had to invest more and more of their resources toward the resolution of these problems, they weren't able to continue expanding like they had previously done. The basic idea is that increasingly complex systems develop new methods for managing problems, but eventually the problems become so difficult and intractable that the systems end up collapsing under the pressure. Tainter introduced the concept of an "energy-complexity spiral" in which societies exploit widely available energy sources to become more complex and solve more problems. But as they become more complex, societies require more energy. Complexity becomes costly, in other words. Tainter does not foresee any wonderful outcomes from these spirals; he's skeptical of everything from resource conservation to technological innovation as methods of addressing major social problems. But Tainter grossly underestimates the ingenuity and adaptability of human beings, and one of the main themes of this work is to express optimism in the powerful potential of human civilization to face and overcome ecological crises through a variety of creative methods.

Other schools of thought have tried to emphasize the chaotic nonlinear dynamics associated with growing social complexity and the prospects of social collapse. Scholars like Gregory Brunk believe that societies are characterized by self-organized criticality (SOC).[33] In this view, societies are supposed to be self-organized systems that periodically return to a critical state hovering on the edge of chaos and stability. In this state of criticality, ordinary social patterns can have cascading effects on the system as a whole, like a small pile of snow that falls and triggers an avalanche. For example, protests occur frequently in any given society, but very rarely do they lead to full-blown revolutions. Every once in a while, however, a sporadic protest spontaneously snowballs into a major social movement with massive political and economic ramifications. In other words, it becomes an avalanche. The Arab Spring is a notable example of this effect in recent history. According to Brunk, human beings have gotten better over time at predicting, anticipating, and addressing the major social problems that we've encountered, which is why "the fundamental reason that civilization has advanced is because societies have become more adept in addressing the problems caused by complexity cascades."[34] One major problem with SOC theory, however, is that it's too focused on the internal dynamics of social systems and less concerned with the external natural conditions in which those dynamics unfold. It's easy enough to wax poetic about complexity cascades and the triumphs of civilization, but only because we've lived through a highly unusual period of climatic stability, the Holocene epoch, that has made civilization possible in the first place. The central concern for civilization is that once those background conditions shift, then the most complex systems are usually the first ones to collapse. That's why bacteria survived the asteroid collision 65 million years ago while some of the most complex creatures that ruled the Earth, like dinosaurs, went extinct.

Earth is a finite place with finite resources, and the same point applies to any other locations in the Universe that humanity might aspire to reach. For all our amazing complexity and achievements, we are subordinate to the natural world, not the other way around.

Leading energy scholar Vaclav Smil wrote that "the planet has finite amounts of elements, it receives and processes a finite amount of energy, and it can support only a finite amount of anthropogenic intervention."[35] In addition, Smil suggests that "the long-term survival of our civilization cannot be assured without setting . . . limits on the planetary scale," further noting that "a fundamental departure from the long-established pattern of maximizing growth and promoting material consumption cannot be delayed by another century and that before 2100 modern civilization will have to make major steps toward ensuring the long-term habitability of its biosphere."[36] This is the basic path we need to follow: setting rational and realistic standards for how to make our civilizations compatible with the surrounding biosphere. But reaching that path will require getting around a few conceptual roadblocks, and none of them is bigger than the dominant narrative about how our economic systems operate.

CHAPTER 2

The Neoclassical World

Before explaining the specific details of the ecodynamic synthesis, we first need to understand a few things about the dominant school of economics that it's meant to replace. The prevailing paradigm of explanation among economists who are supportive of capitalism is known as *neoclassical theory*. Because it's the dominant school, it has great institutional force among governments and corporations, and thus strongly affects how people in power behave in their response to common social problems. Only by dispelling its myths and perversions can we set the stage for another, more comprehensive perspective on how to organize human society in the face of our impending ecological challenges. For much of the nineteenth century, economic debates in Europe unfolded in the context of what's now commonly called *classical economics*, which tried to understand the nature of production by analyzing the role of different productive factors, such as labor and manufactured capital. Classical economists were especially interested in the production process and how it influenced prices and other economic variables. From Adam Smith to Karl Marx, they also emphasized the strong connections between political, institutional, and economic systems. But things began to change in the late nineteenth century, when European economists like William Stanley Jevons, Carl Menger, and Léon Walras established the basis for neoclassical economics by shifting the focus of economic thinking from the analysis of production to an analysis of individual preferences as expressed through exchange markets. This shift made it far easier to obscure the power relations and dynamics involved in creating exchange markets in the first place.

Neoclassical thinkers argued that economic value is subjective and that commodities have no inherent worth, a position known as the *subjective theory of value*. What we are willing to buy in the market, according to them, ultimately comes down to our individual values, to what we find useful and important to us now or in the future. Those values, in turn, are only constrained by the relative scarcities of the goods and services in question. We value things that are scarce and disregard things that are abundant. The neoclassical thinkers claimed that people and businesses want to maximize *utility*, a vague concept that means something like desire. In practice, utility can stand for anything from consumer satisfaction to financial profits, as the situation warrants. According to the theory, economic agents maximize their utility by making decisions at the *margins*, hence supporters of the theory are also known as "marginalists." In other words, when they decide what they want, people do not think about broad categories of goods like "chairs" or "books." They desire a specific chair or a specific book, and then they look at the net costs and benefits that come with consuming that extra commodity. Subject to various constraints, they will keep consuming until they have maximized their utility function, which is another way of saying "until they have fully satisfied their desires." As part of this intellectual program, people were assumed to be rational agents who could make free and independent choices.

The apparent purpose of this neoclassical shift was to make economics a formal science, a seemingly objective field of study that yielded categorical laws and principles about the behavior of human societies. Historians, especially Philip Mirowski, have thoroughly documented how neoclassical theory developed out of flawed analogies and misguided parallels with classical physics.[1] Marginalists reasoned that just as particles are constrained by attractive and repulsive forces, prices and utilities are apparently constrained by the underlying forces that determine supply and demand. In the traditional picture from classical physics, particles respond to the forces they experience in their local environment, like a ball being thrown up in the air and then coming back down under the force of gravity. Forces constrain the dynamical behavior of the particles, and these constraints are described through differential equations with initial conditions. In the same vein, the marginalists believed that prices and wages respond to various forces coming from the real domain. Changes in supply and demand cause changes in prices and wages, almost like a mechanical device with levers, cogs, and wheels. The real domain is analogous to the lever, and the cogs rotate in response to the lever getting pushed and pulled in certain directions.

In his 1892 book, *Mathematical Investigations in the Theory of Value and Prices*, the American economist Irving Fisher described the equilibrium conditions on the marginal utility of an individual and claimed that these conditions correspond "to the mechanical equilibrium of a particle the condition of which is that the component forces along all perpendicular axes should be equal and opposite."[2] He then produced an infamous table in which he made a series of direct and ludicrous analogies between physics and human society. For example, the table seriously suggests that a particle is analogous to an individual person and that energy is analogous to the concept of utility. Whereas energy was subject to *minimization* conditions, utility would be subject to *maximization* conditions instead. But here is one critical difference between physics and economics. Astronomers can tell us precisely when Halley's Comet will return near the Earth's orbit. They can tell us exactly where Mars will be in 56 years. By contrast, economists cannot predict the motion of prices in the same way that astronomers can predict the motion of planets and comets. No one can say with any certainty what the price of red apples sold at the local Whole Foods will be 27 years from now. Prices and wages are not like planets and comets. Because the world of money is a social creation mediated by social relations and institutional hierarchies, it has a tangled web of causation far more complex than anything described by the simple differential equations of classical physics.

Many early neoclassical thinkers used mathematics quite recklessly, making fanciful assumptions simply to obtain a desired result without understanding the full implications of what they were doing. That's the charitable interpretation. The more realistic interpretation is that they simply didn't care about the consequences. These new theorists hoped that economics could be understood as an objective science divorced from politics and the rest of society; that way it could serve as an easy crutch to defend the status quo. The "dismal science" had finally arrived. Once Marxism exploded onto the scene at the turn of the twentieth century, neoclassical theory became the primary intellectual weapon used to defend the class hierarchies and power structures of capitalism, and it retains that role to this very day. However, neoclassical thinkers were gullible to look to physics for their salvation, as if coming up with ill-conceived mathematical equations immediately made their claims more plausible. They should have turned to philosophy instead. Then they would have learned about something called *emergence*, a concept that represents one of the major themes of this entire work. They would have learned that friction and dissipation are important features of our macroscopic world, but they do not exist at the level of a few quantum systems. More to the point: individual cells do not have personalities, but 40 trillion of them interacting the right way constitute human beings who can think, feel, plan, laugh, and cry. People are not particles. The idea that slaves can maximize their utility by improving their marginal position through independent choices is exactly as laughable as it sounds.

Perhaps the central flaw of neoclassical theory is the assumption that our preferences are independent of the rest of society, apart from our interactions in an exchange market. In reality, personal choices are neither subjective nor even that personal—they are causally subsumed in highly complex social networks involving families, friends, co-workers, bosses, political leaders, and numerous other people. Economics is not separable from society and politics. Our preferences and desires do not magically flow from within us—they are shaped and constrained by other people, sometimes in subtle ways that are not immediately perceptible. The implication is that commodity prices and labor wages cannot be understood as microscopic, individual-level phenomena. That expensive shoe is not expensive *just because you really want it*, and your low salary is not the result of you being lazy or dumb. These economic variables are all products of complex social factors converging together across time and space. The people who wield the greatest political, institutional, and economic power are critically important in determining the broad constraints that apply to the prices of certain commodities and the wages of different individuals. The ecodynamic synthesis presented in this work remains sensitive to these crucial facts. Neoclassical economics, on the other hand, is built upon flawed ideas, which even the fanciest mathematical models cannot remedy.

Utility and Prices

The central concepts of neoclassical theory, utility and marginality, both contain a severe defect: they are not empirical quantities that can be measured. Instead of being measured directly, the way someone might measure masses or heights, their values are inferred indirectly from wages and prices, which neoclassical theory assumes provide a window into some hidden world of value. But prices and wages can never fulfill this goal. Commodity prices can change, and often do change, even when the commodities produced are exactly the same. Perhaps these prices are changing because they are measuring some underlying shift in utility, but they could also be changing because of a million other reasons.

There is no way to tell because utility is not something we actually see or measure. In the twentieth century, the economist Paul Samuelson shifted the focus from utility to "revealed preferences," but this concept turned out to be just a more sophisticated phrase that was equivalent to utility.[3] Just like utility, preferences are not measured directly. We are supposed to infer them from observed behaviors. The theory of consumer preferences suffers from other problems too, not least of which is that observed choices do not necessarily result from a constructed set of *personal* preferences. Sometimes you buy that container of ice cream because it was in your "bundle" of preferred goods, but sometimes you buy another container instead because your wife told you to go to the grocery store for her favorite ice cream. The mere fact that you made an economic decision does not necessarily reveal a personal preference. Our choices are embedded in complex social interactions, an important point to remember as we develop our comprehensive critique of the subjective theory of value. Despite its numerous problems, utility is still widely taught in college classes and remains a major background component of economic thought among the marginalists.

In fact, utility hovers over economics like a metaphysical shadow, a Platonic Form that cannot be detected but somehow underlies everything. The major consequence from this philosophical adventurism is that much of neoclassical theory has become ideological, tautological, and nebulous to the point of being incoherent. Even though we can easily point to real-world examples that prove its assumptions wrong, the assumptions are fluid enough that the theory can be made to say just about anything. Consider the assumption that economic agents are "rational" actors that aim to maximize utility. One can quickly dispatch this silly assumption by pointing to philanthropy, among a million other economic activities. If individuals and corporations always wanted to maximize income and profits, respectively, then they would not be giving away large sums of money to charities, even after accounting for sizeable tax deductions or even illegal forms of tax evasion. Neoclassical theorists can always claim, and many do, that people and firms are still pursuing their utility, only in a different way. The inclusion of altruism into mathematical models of utility has become all the rage in certain corners.[4] But these models are largely useless because they exclusively measure altruism through financial transactions. They have no conception of utility and altruism independent of monetary exchange. They have already assumed the truth of what they are trying to show. Notice also that these altruistic models still consider utility to be an individual phenomenon. In other words, when a person makes a donation to a charity, that donation is seen as increasing the individual's utility function. But people can make a donation purely because of social pressure, like when someone tells her spouse to make a donation to her favorite organization. In these cases, which are quite common, the satisfaction does not come from the act of donation, but rather comes from the knowledge of having brought joy to someone whom you really care about. Humans either maximize utility exclusively through self-interested participation in an exchange market or they maximize utility more broadly through various types of social interactions. However, the latter situation requires that we must sometimes decrease the utility we gain from an exchange market, which happens quite often in life but also completely trashes the mathematical formalism of neoclassical economics. The theory is absurd either way, but one can quickly spot how it loves to play chameleon.

Research from the field of behavioral economics has also revealed that people frequently do not act in ways that standard neoclassical theory predicts.[5] People make impulsive financial decisions and succumb to herd mentality. Sometimes they cheat and lie. Sometimes they ignore good evidence and advice, ending up with terrible investments that destroy their hard-won earnings. They are anything but the rational agents that the marginalists need them to be in their useless mathematical contraptions. The marginalists have an answer for this. They will respond by saying that, on average, people do behave rationally over the long run, even if they might do some pretty dumb things on certain occasions. But they have no way of knowing if that's actually true, for the simple reason that things like "rationality" and utility are never measured directly. They are inferred from the same economic variables that they are supposed to explain. Neoclassical theory has no way of showing that people are actually rational; it simply makes that assumption, takes it for granted, and then hopes for the best. Even if you could get past this problem, there would still be an issue with the arbitrary notion of rationality used in economics. One can easily imagine forms of rationality that do not require optimal outcomes in the utility that we derive through exchange markets. For example, we could instead think of rationality as a balance between competing utility functions, some of which describe the pleasure we derive from buying commodities while others describe the pleasure we get from spending time with friends, pursuing our favorite hobbies, and playing games with our children, among other social activities. Sure, this definition is arbitrary, but no less arbitrary than the useless metaphysical abstraction historically adopted by neoclassical economists.

Many of these issues were famously explored by the economist Joan Robinson, who correctly argued that utility is a circular idea. She asserted that "utility is the quality in commodities that makes individuals want to buy them, and the fact that individuals want to buy commodities shows that they have utility."[6] One cannot infer utility from prices and then shamelessly proceed to infer prices from utility. The implication is, once again, that prices cannot offer us an exclusive window into individual values. Commodity prices can change because of many reasons, from political elections to violent wars and revolutions. Trying to isolate the influence of individual "values" among all of these factors is an absolutely hopeless task because so many of these factors are integrated together. People could like an article of clothing for its color, fabric, artistic value, and a million other reasons. For example, a t-shirt can become instantly popular one day because a celebrity was seen wearing it and got photographed by the paparazzi. Some commodities, or types of commodities, can gain visibility in the public eye through highly chaotic and unpredictable ways. Product endorsements are also common nowadays, especially with social media personalities who sometimes create their own product lines, be it for cosmetics, clothes, food, and many other things. Economists do not consider the importance of social influence at all when trying to understand price dynamics. That omission makes sense for their institutional role within the capitalist system; they have settled for the usual propaganda that prices represent objective and reflexive signals from the "real" domain of physical production. Saying that society might play a role in setting prices sounds too subjective and irrational for those in charge, who need people to believe that the economic realm is a perfect reflection of fundamental scientific laws, instead of an elaborate construction that benefits those who have wealth and power.

Supply and Demand

It's certainly true that supply and demand provide important *constraints* on economic activity and the dynamics of the nominal domain. If 95 percent of the world's apples suddenly disappear tomorrow, it's a safe bet that global apple prices will rise dramatically. In the critique that follows, I am *not* rejecting supply and demand as useful categories of economic analysis; I am *specifically rejecting the way that these two concepts have been abused and manipulated within neoclassical theory itself.* I will happily employ supply and demand as explanatory concepts later in the book, but only by placing them within broader causal and historical contexts. It's absolutely misleading to believe the naïve Newtonian picture in which supply and demand function like mechanical forces that pull and push prices in different directions. Too often in both academic and public discourse, it's this silly picture that's blindly repeated, and virtually all changes in prices, wages, and profits are automatically attributed to "supply and demand" like a Pavlovian reflex. This was especially the rationale du jour during the height of the recent global pandemic, when much of the media and many academic economists kept spitting out the phrase "supply and demand" for any unusual patterns in consumer goods prices, labor markets, real estate markets, or any other strange activity in the global economy, making hardly any further attempts at deeper explanations.[7] We've arrived at a point where "supply and demand" has become a popular catchphrase, repeated just because everyone else seems to be saying it, but without any actual awareness of its meaning or implications. In reality, the nominal domain of prices, profits, and wages is heavily filtered and structured by intermediary social relations and various economic and political struggles over the distribution of economic resources.[8] Prices don't just move in tandem with changes in the biophysical world; their actual trajectories are heavily influenced by intervening social dynamics.

Let's try to better understand why this seductive catchphrase is profoundly misguided. The traditional neoclassical story tells us that prices are determined by supply and demand. The law of demand states that, all else being equal, prices are inversely related to the quantities demanded. If something becomes more expensive, people will want to purchase less of it. And the law of supply says almost the opposite: all else being equal, prices are directly related to the quantities supplied. If prices rise for a particular commodity, then producers will want to supply more of it, so they can make more money. The dynamic competition between these two economic levers is supposed to yield the "equilibrium" price. That's the neoclassical story in a nutshell: scarce things are expensive and plentiful things are cheap. But this insight is largely vacuous for many reasons.

First, let's consider the problem of causation. If changes in prices are caused by changes in supply and demand, then what's causing changes in supply and demand? One cannot fall back on tropes about how there's a feedback loop here, because real feedback loops are always causally embedded in a larger environment; they're never isolated. For example, a major theme of this book is the energetic feedback loops between human society and the natural world. But these feedback loops can only function in the context of a causal arena featuring light energy from the Sun, gravity from the Earth, and many other necessary biophysical factors. It's therefore not convincing to rely on circular reasoning and to claim that prices and supply and demand cause each other. The basic truth is that supply and demand are incidental causes, at best. Price dynamics are fundamentally caused by social and political struggles over the distribution of economic resources. Supply and demand cycles are usually engineered by powerful individuals, corporations, and governments, as a strategy to yield the prices and profits they ultimately desire. Gas prices in the Western world skyrocketed in 1974 because the OPEC cartel temporarily suspended oil shipments to Western countries as a way to punish them for supporting Israel in the October War; the price of gas didn't rise because a shortage magically and spontaneously appeared out of nowhere. It increased because of geopolitical dynamics, which caused the shortage in the first place. For another example, the De Beers cartel deliberately restricted the supply of diamonds in global markets throughout the twentieth century, leading to grotesque price inflation along the way.[9] A typical neoclassical economist would look at the situation and blame a supply shortage for the high price of diamonds, just like they'd reflexively blame a supply shortage for the high price of anything. But there was only a supply shortage because De Beers *artificially created one* to cement its stranglehold over the diamond industry. To think of supply and demand as omnipresent forces that somehow control prices behind the scenes is to ignore the active agency of powerful individuals and corporations in establishing the critical parameters of the nominal domain, from prices and wages to profits and interest rates. Imagine a child hitting a baseball that accidentally smashes the neighbor's window and then has the nerve to tell his neighbor, "Well I didn't break the window. *The ball did.*" In a silly and pedantic way, it's of course true that the ball physically penetrating the window caused it to be smashed into pieces. But then again, it was the child who swung the bat and gave the ball its unfortunate trajectory, and this is the cause we ultimately care about. Neoclassical economists recycle ritualistic propaganda about supply and demand and the "invisible hand" of the market as a way of obscuring and marginalizing the critical factors, like social power and class domination, that collectively have a far more profound impact on the dynamics of the nominal domain.

And then there's the problem of demarcation. It's not easy to know how to define or measure supply and demand in specific circumstances. Take oil as an example. What is the global supply of oil? Is it all the oil on planet Earth? Is it all the oil in proven reserves? All the oil in commercial or strategic inventories? How about the finished oil products stored in refineries or marine terminals? Take housing as another example. What's the supply of housing? Is it the stock of existing homes for sale? The stock of new homes for sale? Both combined? What about non-rental vacant properties? What about new finished homes held back for inventory? The fundamental reason why this issue matters is because we might find results that are consistent or inconsistent with the laws of supply and demand depending on how we define these terms and what we actually measure. If the global oil supply is all the untapped reserves on planet Earth, then oil extraction over time depletes that finite stock, implying that prices should get consistently and continuously higher, if the law of supply is right. But that's not what we see; oil prices actually exhibit huge swings and variations over time. Given these problems, economists typically understand the term *supply* to mean whatever is available for sale in a market. Business economists often differentiate between the *stock*, which is what the seller holds in reserve, versus the supply, which is the amount offered for sale in the market.[10] This distinction is relevant because market supply is often poorly defined, and the available supply will depend on what we decide to include as part of the market in question. But more important, supply is often artificially constructed by those who have the power to do so. Capitalists routinely create artificial scarcity by controlling how many products show up in markets in the first place. As mentioned above, De Beers used to keep diamonds off the market so it could inflate diamond prices, and OPEC does the same thing with its production quotas on oil. Ticketmaster often sequesters a large percentage of tickets for major events to justify higher prices on the remaining tickets that it does offer for sale to consumers. Real estate developers keep many of their finished homes from getting on the market, precisely to justify higher prices on the homes that are available for sale. Pharmaceutical companies in the United States use a variety of legal and political methods to limit available drug supplies, thus pumping up prices for millions of people. Restricting supply is a time-honored capitalist tradition designed to make goods and services seem like they're special and exclusive. For that reason, defining supply as just the stuff that's available for sale in the market is often an ideological cover for the corrupt and self-interested decisions of elite capitalists.

Demand is even more notoriously difficult to define than supply. Here's how the economist Susan Feigenbaum puts it: "The quantity of a good a person is willing and able to purchase at any given price during a specified time period . . . *all other factors held constant*."[11] Two other economists used a bit of a mathematical caricature to define the concept: "Demand = Desire to Acquire + Willingness to Pay + Ability to Pay."[12] These definitions all sound great and intuitive, but it doesn't take much thought to realize that they're pseudoscientific pablum. How does an economist exactly measure someone's desire or willingness to purchase something? The short answer is that they can't; all they can do is measure what products were purchased, in what quantities, and the corresponding prices. Desire and willingness are neurobiological phenomena that cannot be accurately and consistently measured with our current technological systems. Speaking in these silly terms is part of the lazy neoclassical effort at explanations based on methodological individualism, the notion that economic behaviors and decisions are derived from the autonomous preferences that exist within individuals. But the reality is that people take plenty of economic actions that have absolutely nothing to do with their internal desires and personal preferences, simply because we're social creatures whose actions are often influenced, and sometimes even forced or manipulated, by others.

As if these issues aren't bad enough, there's also the problem of time. It's possible to get conflicting conclusions about the laws of supply and demand depending on the time intervals under analysis. This issue points to a general problem in economics: a cherished empirical relation might hold for five years or so, then break down completely once we get out to twenty or thirty years. Alternatively, the relation in question might hold up pretty well over long periods of time, but could easily break down over short time intervals, such as weeks or months. Another issue is the problem of stability and equilibrium. For any given price point, there exist an infinite number of supply and demand curves that could generate that price. Because it's practically impossible to measure supply and demand, at least as they're typically defined, it's also practically impossible to identify which curves are responsible for generating any given feature of the nominal domain, from individual salaries to commodity prices. This is an especially important point, because even if we grant the neoclassicals all their wildest fantasies about supply and demand determining everything, the practical consequences of that admission are virtually irrelevant as there's no way to empirically nail down the supply and demand curves to which specific markets are supposedly responding. And if we can't do that, it becomes hard to make policy recommendations based on supply and demand considerations, since we wouldn't know which supply and demand curves apply to households and businesses, both at a microeconomic level and at an aggregate level.

In the 1970s, the economists Hugo Sonnennschein, Rolf Mantel, and Gérard Debreu published a series of papers concerning the uniqueness and stability of general equilibrium in neoclassical economics.[13] General equilibrium is a hypothetical macroeconomic state where aggregate supply is supposed to equal aggregate demand. Their work came in the context of earlier results from Debreu and the American economist Kenneth Arrow showing that general equilibrium could exist, but only under highly idealized assumptions that apply absolutely nowhere in the real world. The results of Sonnennschein, Mantel, and Debreu collectively became known as the "SMD theorem" after their last names. The SMD theorem states that general equilibrium, even if it exists, is neither stable nor unique. If an economy reaches a state of general equilibrium, it won't be able to stay there. And what's worse, there are multiple paths toward general equilibrium, opening up the problem of which path we should pursue. In short, the SMD theorem is a highly negative and deflationary result for neoclassical theory because it shows that even if you know the equilibrium prices that prevail in general equilibrium, that information cannot tell you anything about the underlying economy that actually produced those prices. In effect, there are many "microscopic configurations" that can produce the same state of general equilibrium. Later results from economists like Alan Kirman, Donald Saari, Donald Brown, Chris Shannon, and others have only strengthened and expanded the original conclusion.[14]

Finally, there's the problem of interdependence. Neoclassical economists often think that supply and demand are autonomous and independent forces that jointly determine economic outcomes. That's the fantasy. In the real world, however, supply and demand are not actually independent functions that magically settle on some equilibrium price. The two are fundamentally synergistic and interdependent precisely because governments and corporations usually try to control both levers when they plan for the future. Large corporations actively intervene to shape and influence the demand patterns of consumers through a bewildering array of strategies. They spend hundreds of billions every year on advertising to persuade people to buy junk they don't need. They also spend vast sums of money to lobby politicians, to make campaign contributions, and in some cases to outright bribe lawmakers to get their desired legislative result. They exploit their monopoly power to knock out potential competitors and to erect barriers to market entry, thereby substantially narrowing and limiting the available market options for consumers. This speaks to a broader truth about capitalism: markets never bring about order by themselves, they are always constructed from preexisting orders and social structures. Dominant corporations don't just sit around waiting until consumers "like" their products; they ruthlessly manipulate the law and the political system to shape and corner the market as they wish. The notion that economic outcomes are the product of decentralized and distributed networks is a lazy and vapid fairy tale designed to excuse the failures of the status quo.

All social and economic orders are constructed from the dynamic interplay of preexisting class and power relations. But elites and capitalists are largely missing from neoclassical theories. In this fantasy world, it's only consumers and regular people who make choices, like what house to purchase, where to go to college, or what business to start. The rich and the powerful apparently have no agency whatsoever. It's not capitalists who set wages and prices; it's the market, through spontaneous orders, invisible hands, and ghosts in the machine. In the neoclassical paradigm, the market has the same function as God did for elites in the Middle Ages: it's a convenient deus ex machina for justifying and eternalizing the current structure of the world. In the neoclassical vision, the consequential decisions of powerful elites are reimagined and abstracted as mysterious market forces, both to legitimize the social impacts of those decisions and to obscure their true origins, making the resulting social order seem completely normal and natural instead of imposed and constructed.

In practice, it's also quite difficult to empirically test most claims about supply and demand. Take demand as an example. Sure, it may be easy enough to notice simple correlations between rising prices and lower sales. But the "law" of demand has another part, its infamous *ceteris paribus* condition, "all else being equal." This condition implies that quantities can change for *many other reasons* besides price fluctuations, and vice versa. The empirical problem for the economist is to show that any change along the demand curve was specifically caused by variations in prices, as opposed to the thousands of other factors that could have driven the *exact same observation*. But how exactly is one supposed to do that? How do you keep "all else" equal in any real-world situation? How does one account for all possible confounding factors and variables? There are numerous statistical methods in econometrics to deal with these kinds of issues, but absolutely none of them are foolproof, and neoclassical economists are infamous for creating juvenile models that explain nothing useful about the world. Unlike laws in the natural sciences, most of the claimed "laws" in economics are pseudoscientific gimmicks meant to rationalize the existing order.

The ultimate constraints on supply always come from nature, but society can and does intervene in numerous ways to adjust the levels of supply and demand that are actually available. This intervention can occur through concerted government action, through strategic economic decisions from dominant groups, through various kinds of class struggles, or through some dynamic combination of all these things. The conservative British prime minister Margaret Thatcher once seriously suggested, in an effort to deprioritize the role of government in solving social problems, "There is no such thing as society. There are individual men and women, and there are families."[15] This is a bit like saying, "There is no such thing as a human being; there are only atoms and molecules." And if you want to explain and understand human behavior, you must do so at the level of atomic and molecular interactions. Of course, even atoms and molecules are not the most basic constituents of nature. We can just keep the reduction going all the way through to strings and branes, if we care about consistency, assuming that those things exist. In parroting this nonsense, Thatcher failed to grasp that what constitutes a "thing" is not just its components, but also the *interactions that underlie the components*, which is why the "thing" is capable of changing in the first place. Individuals do not exist in a vacuum, sunbathing on their private islands away from everyone else. Society represents the mental abstractions and organized interactions that allow people to communicate and work together. It does exist; it does affect the distribution of goods and services to individuals; and it does provide an extra lever of constraint on supply and demand. A glib statement like "prices are determined by supply and demand" means nothing unless it considers the role that various social forces play in establishing and reinforcing those cycles. In the words of economist Mariana Mazzucato, "Prices and wages are often set by the powerful and paid by the weak."[16] Neoclassical economists largely ignore these hard realities, pretending that individual preferences somehow come out of thin air, divorced from the causal webs and structures of society and nature.

Production and Distribution

Although the concept of marginality permeates every feature of neoclassical theory, it plays an especially important role in the context of productivity and the neoclassical theory of distribution. The original goal of this theory was largely political: to defend the power of capital by telling workers that they lived in a fair and just economic system. The marginalists argued that economic agents obtain returns that are equal to their marginal products, assuming certain economic conditions. The marginal product of an input is the net gain in productive output that comes from the addition of an extra unit of that input to the productive process. This definition implies that workers earn salaries corresponding to their company's net increase in productivity. Be more productive and you get a higher salary. Similarly, businesses earn profits that equal the net value of output they produce for society. In his influential 1899 book, *Distribution of Wealth*, the neoclassical economist John Bates Clark wrote: "It is the purpose of this work to show that the distribution of the income of society is controlled by a natural law, and that this law, if it worked without friction, would give to every agent of production the amount of wealth which that agent creates."[17] Clark continued by plainly revealing the ideological purpose of his work: "The welfare of the laboring classes depends on whether they get much or little; but their attitude toward other classes . . . depends chiefly on the question, whether the amount they get . . . is what they produce. . . . If it were to appear that they produce an ample amount and get only a part of it, many of them would become revolutionists, and all would have the right to do so."[18] Translation: let's try to justify the income that flows to capitalists by making them appear to be productive.

An immediate problem with this fairy tale was that Clark and the marginalists had no objective way of measuring marginal productivity. What is "productive output" exactly? Economists had absolutely no clue, and they still don't. They simply assume that a company's sales are equivalent to its productivity, when in reality sales are just a company's income, which can be caused by many different and complex factors. The problem with the glib notion that sales are the same thing as productive output is that it eliminates the possibility of explaining those sales through some independent measure of production, because the neoclassicals *artificially defined sales to be productivity*. They fell into a circular trap: the marginalists had no idea how to think of productivity *independently* of things like profits and wages. Instead of using productivity to explain sales, all neoclassical theory does is simply define productivity *in terms of sales*, then turns around and calls that an explanation. But if productivity is defined in terms of the nominal domain, then it *cannot be used as a causal explanation* for observations in that very same nominal domain. If X is the explanatory variable for Y, then we cannot use Y to explain X, because we're saying nothing more than *Y explains itself*, which is absurd. Neoclassical economists might argue that productive outputs are explained by productive inputs like capital and labor. But this rebuttal gets them nowhere, for how are these inputs actually measured by economists? Once again, they're practically measured in financial terms, the very same financial values that an independent notion of productivity is supposed to explain in the first place. In blindly equating productivity to monetary values, the neoclassicals lost the ability to use the concept of productivity as a way of explaining the dynamics of those monetary values. An easy analogy here is to medicine. When someone gets a fever, we all understand that the fever is a *symptom* of some underlying disease; it's not the cause of anything by itself. The cause is the virus inside the body and the fever is a biological manifestation of the body fighting off the virus. In medicine, there's an obvious difference between the cause and the symptom, *and they can both be measured separately*. One can measure body temperature and notice that it's high, indicating a fever. One can also take a lab test and confirm the presence of the underlying virus. Everything in this example is conceptually simple and straightforward. But in neoclassical economics, that critical distinction between cause and effect is absolutely destroyed, to the point where the cause, productivity, and the symptom, the nominal domain of financial sales and incomes, are practically and artificially defined to be the same thing, leading to a vacuous theory devoid of any meaning.

There are also several empirical and observational issues that strongly refute the neoclassical story. From standardized test scores to observational studies in applied psychology, an overwhelming amount of empirical evidence indicates that human ability and productivity are normally distributed.[19] They follow the familiar "bell curve" distribution common to so many other random variables, such as height and weight. However, the same thing is not true for income and wealth. It turns out that these generally follow a power-law distribution. These distributions are characterized by a minority of extreme values, the "long tails" that stretch far beyond the normal range of data. The details of these statistical distributions are not important for our purposes. Here's the only thing that matters: variations in wealth and income are far greater than variations in ability and productivity. We are forced to conclude that productivity alone cannot successfully explain the distribution of wealth and income. If salary differences reflected productivity differences, then salaries would be more or less normally distributed as well. Some very important details are obviously missing from the neoclassical theory of distribution.

Another major problem is that productivity is not simply an individual trait. It's a social effort. In actual work environments, people have to interact and communicate with other people. Workers are not isolated towers; they are embedded in certain social and productive relations in the workplace. This interconnectivity makes it almost impossible, in practice, to disentangle the productivity of one worker from another, regardless of how we decide to measure productivity. Numerous research studies in applied psychology have revealed the fundamental importance of social familiarity and coordination on team performance and productivity.[20] Finally, consider the aggregation problem, that recurring nightmare for neoclassical theory. In the real world, people do many different kinds of jobs. Some people work as machine operators and retail clerks, others work as insurance agents and senior marketing directors. There is no obvious unit of measurement that would apply equally well to the productivity of all these people. How do you compare the physical output of a farmer to that of a lawyer? Remember that we cannot use profits and wages because those are the *very things we want to explain*. The fundamental point here is that neoclassical theory has absolutely no clue how to even define productivity, separately from monetary exchanges.

Neoclassical theory has a strong fixation on justifying the profits that flow to capitalists. One of the primary ways it does so is by arguing that capital is productive, hence the owners of things like factories and machines deserve to benefit from their productivity. An immediate objection, as Marxists would point out, is that factories and machines are themselves built by human labor, so they're only productive in a secondary and derivative sense. It's for this reason that Marx called capital factors "dead labor." Beyond this critique, there are also severe methodological flaws in how neoclassical economists measure and understand the concept of capital. In macroeconomics, we can supposedly measure "real" output through mathematical constructs known as *aggregate production functions*. An aggregate production function takes aggregated inputs, like labor and capital, and gives out the maximum possible output that those inputs can generate. There is no problem with this procedure as a general abstraction. The problems arrive specifically when the inputs and the outputs are measured in monetary terms, using dollars or some other currency. These are exactly the kinds of games that economists play in practice, despite the futility of trying to aggregate using commodity prices and distributions that diverge in highly chaotic ways. Virtually all financial aggregates that appear in macroeconomics are simply mathematical artifacts that have no concrete meaning. Let's consider the problem of aggregating capital through monetary units. In an economy with different kinds of commodities, it does not seem obvious which units we should pick to add up things like apples, computers, and chairs. One could pick stable and reliable units, like kilograms for mass or joules for energy, but that would be too rational and scientific. Economists instead wax poetic about the "capital stock," even though no one really knows what that means.

Textbooks pretend that the capital stock is a collection of productive factors, such as tools, equipment, and machinery. This definition is wrapped up in the ideological desire to see capital factors as productive, thereby justifying the profit rates obtained by the capitalists. But here's the basic reason why this approach leads nowhere. How does one compare the productivity of an office computer to that of a robotic arm operating in a factory? Capital goods are used for different purposes; they perform different actions and produce different things. They're "heterogenous," to use the official term among academics. We're faced with the problem of coming up with a common unit of measurement for heterogenous capital stocks, and most economists have no better answer than resorting to the nominal domain, using things like prices, profits, and sales as proxies for productivity. Economists ignore natural units when measuring the productivity of capital goods; they pretend to measure productivity using monetary values instead. This brings up an immediate problem. How do we measure monetary values for capital without invoking the rate of profit, which is what the valuation of capital is supposed to explain in the first place? If capital is productive and helps to explain why firms score profits, then profits cannot be used as a proxy for productivity, otherwise they would be self-caused. The value of the capital stock depends on profit rates, but profit rates themselves depend on the value of the capital stock. We are stuck in a circular loop, as many economists have recognized over the years, including the likes of Knut Wicksell, Joan Robinson, and Piero Sraffa.

To get a concrete sense of this problem, consider an oil company that wants to figure out the total financial value of its oil tankers. It assigns an average price per tanker and then multiplies the total number of tankers in its inventory by that average price. Easy, right? Not so fast: oil prices fluctuate up and down, sometimes dramatically so. The company's profit rate, along with its market value, may decline if oil prices fall sharply. And the profit rate of the company affects the sale price of the tankers. When oil is expensive, the tankers are worth more because they're transporting an expensive commodity around. They have precious cargo on board. But when oil becomes very cheap, the tankers are generally worth less than before. Let's be clear about the fundamental problem here. The monetary valuation of the tankers allegedly serves as a proxy for their productivity, so this monetary metric of productivity is supposed to explain the profit rate of the company. However, it looks like the profit rate of the company is actually explaining the valuation of the tankers! In reality, it's even more complicated than that because the company's profit rate also depends on the financial valuation of its tanker fleet; it's a two-way street. An oil company that can charge more money when it sells a tanker will obtain higher revenues, and potentially higher profits. But there's also a flipside to this story. If the tankers are valued more "on the books," their insurance costs might be higher, potentially hurting the profitability of the company. The central point is that the "value" of a firm's capital stock is integrated with the company's profit rate in highly complex ways, and the profit rate itself depends on the valuation of the capital stock.

This inescapable reality makes it extremely difficult, if not impossible, to measure the physical and "productive value" of capital in monetary terms, if we're using the term "productive value" as a reference for the source of the profit rate. The price of capital factors is highly variable and unstable, reflecting the chaotic conditions of economic activity. When it comes to capital as an economic unit of analysis, most economists suffer from severe schizophrenia. They have defined capital as everything from a physical factor of production, like factories and vehicles, to the amount of money that a business or an individual has available for investment. Other uses and definitions are also found in the literature, suggesting that the concept is far too cryptic and ambiguous, to the point where it means virtually nothing. In refusing to specify a scientific unit for measuring capital, economists have deliberately left the concept as a vague black hole that can satisfy every wish list. Sometimes it can be physical stuff that we use in production cycles. Sometimes it can be profit. Sometimes it can be net worth. Capital is only limited by our imaginations.

Time and Money

If harping about capitalists being productive doesn't quite explain their profits, the marginalists made sure to have another trick up their sleeve. The beauty of having an ill-defined and ambiguous theory is that the potential for rubbish is virtually endless. Because the marginalists had no idea what determines the rate of profit, they made sure to offer several potential answers. That way at least one column would remain standing if the others ever collapsed. One of those answers, the theory of "time preference," originated with the Austrian economist Carl Menger and received further refinements from his compatriot Eugen von Böhm-Bawerk and the American Irving Fisher. The theory goes roughly like this: People value the consumption of goods and services in the present more than they value the consumption of the same goods and services in the future, hence the time preference. In order to delay the instant gratification that comes with consumption, people must either pay a price or earn a reward. The price that we pay for our time preference is interest, and the more we value consumption in the present, the more likely we are to pay a higher rate of interest when taking out a loan. From the perspective of the lender, the situation is reversed. The lender already has lots of money. He could spend that money right now on things like consumer goods and vacations. But he could also invest it. Investing the money now means that the lender is delaying his gratification in hopes of earning a bigger reward down the road.

The reward for that delayed gratification is the interest rate that the lender demands for his investment. Different people can have different time preferences. Some might really want lots of money to spend right now, perhaps to buy a big house or something. They will usually pay high interest rates for that preference. Others might want to invest lots of money in the present, perhaps by buying lots of stocks. They will typically expect high returns for doing so. But the basic point is this: interest rates are the costs associated with the passage of time. In this view, interest rates reflect our time preferences, not the productivity of capital. Piero Sraffa may have axed the idea of capital as an independent factor of production, but neoclassical theory still had other ways of explaining profit rates. In the 1884 work *History and Critique of Interest Theories*, Böhm-Bawerk criticized socialists for ignoring the role of time in the production process. He reasoned as follows. If all commodities are produced instantly, then capitalists could never score any profits because there would be no risks and uncertainties involved. But production is distributed across time. Certain goods are produced in the short term, then those goods are used to produce other goods in the medium term, and then those combined goods produce other final goods in the long term. Profit is the necessary price for this "roundabout" process.

Böhm-Bawerk claimed to agree with socialists that workers should be paid the full value of their labor, but he wanted to clarify what that means by distinguishing between *present* value and *future* value.[21] He maintained that workers can either be paid the present value of their labor right *now* or they can be paid the future value of their labor in the *future*. But they should never be paid the future value of their work right now, which is what the socialists were suggesting, according to him. This line of thinking requires money to have a certain *time value*: a dollar today is worth more than a dollar a year from now because the current dollar can be invested and can start accruing some interest. Here is where Böhm-Bawerk made his critical point: because it takes time to produce commodities, the workers receive their wages *before* the final products they made are sold in the market. What this implies is that a time difference exists between the compensation of the workers and the sale of the commodities. This time difference is the reason why capitalists score a profit, according to Böhm-Bawerk. Capitalists make more money from the sale of commodities to consumers than the amount they are paying to their workers because the workers are paid before the commodities are sold. In effect, the salaries of the workers are discounted to their present value from some expected future output. The workers cannot be paid the full value of what they have not yet completed, and so the capitalist takes home the difference between the full value of the product and the value of what the workers have done until now. In this view, profit is a natural part of production, not a consequence of greedy confiscation by capitalists.

Böhm-Bawerk pads his argument further with an example. He considers a worker making a steam engine. If the engine requires five years of work to manufacture and has a final price of $5,500, then the worker who produced the engine should receive $5,500, assuming he worked all five years on the engine. But suppose the worker left the job after only a year. Böhm-Bawerk asks us to think again about what wage he deserves now. One response might be that he should get $1,100, which is one-fifth of $5,500. However, Böhm-Bawerk argues that this position is wrong because $1,100 is one-fifth the price of a fully completed steam engine, which is not what the worker has produced. Instead, the worker has produced some unfinished component of what could become a steam engine four years from now. Because present goods are worth more than future goods, one-fifth of a fully finished steam engine right now should have a higher value than one-fifth of a steam engine that won't exist for another four years. The worker should thus be paid less than $1,100 if he wants the money after one year and not after five years, when the completed steam engine can be sold. But how much less? Now we enter the realm of subjectivity; the amount of the worker's salary depends on the discount rate we pick. If we discount the market value of one-fifth of a steam engine at a rate of 2 percent over 5 years, then the worker would be paid about $1,000. If we pick a discount rate of 7 percent instead, then the worker would get about $800, which is obviously much less money.

But here's a simple question that could tarnish the immaculate fantasies of the Austrians: Who decides the discount rate? Who decides how much to discount the expected future value of the commodities produced by the workers? The capitalist does, of course. In effect, the capitalist establishes both the interest rate and the discount rate in one fell swoop, and he can pick whatever he wants. Nothing in this argument constrains his choice, apart from the condition that the worker's wage should be higher than $0, otherwise the worker would be little more than a slave. The Austrians could claim that the interest rate would constrain the discount rate because the two are the same thing. But nobody knows what the interest rates will be for commodities that are sold five years from now. The capitalist knows the numbers for prior interest rates and may try to infer the future ones from those, but there is no guarantee that this hopeful extrapolation will actually materialize. The Austrians might retreat to time preference: the desire for profits now is what sets the interest rate, and the discount rate by extension. Now we have a circular quagmire. Time preference fixes the profit rate and then the profit rate establishes the time preference. The capitalists basically decide their own profit rates. This classic argument by Böhm-Bawerk, truly a masterpiece in flawed reasoning and grandiose twaddle, has made the rounds quite often in the history of neoclassical thought. But it does not invalidate the idea that capitalists exploit their workers. It simply provides new constraints on the space in which that exploitation can happen, assuming that its premises are correct. Of course, its fundamental premises are not correct, and now we can finally proceed to tear them apart.

Böhm-Bawerk picked a very convenient example for himself: a situation with a large time difference between the worker's compensation and the sale of the commodity. But suppose we have a case with a much smaller time difference, which is certainly the case for most products that are not as complicated as a steam engine. Even with a marginal time difference, capitalists routinely score huge profit rates. This is true even once you take into account the roundabout nature of production, the fact that several stages of labor and goods are involved in the delivery of a final commodity for exchange. Production cycles with quick turnarounds are not only capable of delivering huge profits, they are the fundamental basis of those profits. As we're going to see, a major economic goal of technological development is to speed up dynamical cycles by delivering more services and commodities in a given unit of time. Faster dynamical cycles coupled to a liquid monetary environment generate larger networks of circulation and distribution. The general rate of profit then represents a changing order parameter that tracks how much influence dominant capitalists have over the dynamical and distributional cycles in the economy. Contra Böhm-Bawerk, capitalists do not derive their value from the passage of time. They derive value and power from financial accumulation, from the fact that they strive to control and accumulate more capital than other peer competitors. It follows that time difference cannot be an important factor in the determination of interest rates. Money has absolutely no *inherent* time value. The idea that long-term profit rates should be positive is simply a transitory and historically contingent phenomenon of modern capitalism, incessantly parroted by the rich and ingrained in society more broadly as a pervasive cultural expectation.

Nor does Böhm-Bawerk's argument translate neatly into our contemporary age of financial capitalism, where the time differences involved could be a few minutes, much less months or years. Consider a stockbroker who must convince clients to keep buying stocks. For every successful transaction, the broker earns a commission fee. But she does not keep the entire commission; she must split it with the brokerage firm. However, the present value of her labor is the full commission, which is being delivered right now, not five years later. If she deserves the present value of her work, she should get the entire commission, according to a straightforward reading of Böhm-Bawerk. His argument also has nothing useful to say about arbitrage, the buying and selling of assets for the purpose of exploiting a price difference between two markets. A high-frequency trader who buys an asset at one price and then sells it in a different market for a slightly higher price, all in the fraction of a second, can make a huge profit while incurring virtually no risk during the transaction. High-frequency trading is now a major component of the modern financial system, making up to a quarter of all trading volumes according to some studies, if not more.[22] The neoclassical fantasy genre claims that arbitrage opportunities should not exist or should disappear very quickly. But not only do they exist, they proliferate widely in currency markets across the world and deliver consistent returns spanning decades, so clearly some capitalists are making huge profits without batting an eyelash about risks and time preferences.[23]

Finally, consider some major structural features of modern financial systems in the Western world and how they combine to reduce or eliminate risks. All member banks of the Federal Reserve System used to obtain a virtually *guaranteed* 6 percent annual dividend on their invested assets in the Federal Reserve, the central bank of the United States.[24] Smaller financial institutions are still receiving the dividend to this day. I say "virtually" guaranteed because the Federal Reserve must score a profit to hand out those dividends, but "earning" this profit is a mere formality since the Federal Reserve manages a large portion of the money supply for the entire country. This brings up an even more fundamental point: the biggest banks and lenders in the Western financial system, from JPMorgan to BNP Paribas, can practically do whatever they want without incurring any *systemic* risks because of their strategic alliances with the state. If some shocking event happens and threatens their very existence, the government has and will always intervene to save them from collapsing, given their importance to the wider economy. Citigroup has been bailed out by the United States government no less than three times in the past century. Risk and time preference have nothing to do with the handsome profits of the major private banks; these profits largely derive from their fortunate strategic position in the wider class and political architecture of modern capitalism.

At this point, the marginalists may decide that other factors besides time are involved in setting interest rates. But whatever reductionist approach they choose will also explain very little in the end, because the core problem is not with their arguments, but with the assumptions behind their arguments. When the marginalists weaved together a simple tale about returns and productivity, people like Robinson and Sraffa could at least refute their bedtime stories on the basis of mathematical reasoning. But when it comes to time preference, the issues involved are far more nuanced and philosophical. As an offshoot of the subjective theory of value, time preference suffers from the same problems. The marginalists claim that multiple interest rates can exist for the same economic process because people have different time preferences. That leaves open the problem of explaining *why* people have different time preferences. The subjective theory of value turned out to provide no useful basis for understanding economic behavior, even as an emergent explanation of the world, and the concept of time preference does not fare any better. Consider the most basic claim of the theory: we value consumption in the present more than consumption in the future. This claim implicitly assumes that we have complete information about the future, otherwise we could never know how much we value future consumption. But no one has complete information about the future, and thus no one really knows how much they would value future consumption. They can certainly estimate and anticipate future levels of investment and consumption, but this assertion is an entirely different one than the claim about what people actually value at different points in time. We simply have no idea how much value we would attach to consuming a slice of pizza next year versus having one right now. Maybe some unique life event intervenes next year and makes you value the consumption of pizza even more. But you would never know anything about that right now.

The Austrians could respond by saying that people value the certainty of the present over the uncertainty of the future. You may not even live to see next year, so it's better to have that pizza now instead of counting on it later. This claim is certainly true for many people, but it still does not explain the degree to which we value certainty over risk and uncertainty. It does not explain the vast divergences in the way that people order their time preferences, in the way that they consider, tolerate, and understand the nature of risk. The marginalists fail to grasp that our time preferences are coupled to our social expectations and economic realities. Consider how the United States government prevented the collapse of so many major companies during the Great Recession, including heavyweights like AIG and General Motors. During the coronavirus pandemic in 2020, the federal government essentially bailed out *entire industries*, such as airlines and hotels, by pumping them with cash and extending generous loans. A capitalist who has managed to secure generous government subsidies for his business will have a very different perception of risk than many of his counterparts, who may not have been that lucky. But it would then be the height of folly to attribute the profit rates of this capitalist

THE NEOCLASSICAL WORLD

exclusively to his time preference for investment, which depended heavily on external factors all along. Again, these examples are not some silly abstractions; they happen frequently in capitalist economies and have a huge impact on the bottom line. Private capital in the West seems to be on the verge of collapse about once every decade, and direct intervention from the state is the only thing that saves it.

Böhm-Bawerk said that capitalists deserve their profits now because the production process takes place over time. But there are other complex dimensions to the production process besides time. Production certainly unfolds in time, but it also unfolds in space and becomes embedded in the particular social and economic relations that prevail within that space. A capitalist who wants to shutter a plant in the United States and open a new factory in a place like India or Vietnam, where labor costs are lower and the governments hostile to workplace dissent, immediately knows that a wide array of complex social and political factors are about to boost the profit rate. The foreign government trying to entice the capitalist to move his business could even offer numerous kinds of financial bribes and sweeteners to seal the deal. The overall effect for the capitalist is that his investments now face a much lower risk profile. In other words, the idea that capitalists deserve whatever profits they get because they are taking huge risks with their money is an absurdity, precisely because dominant capitalists, through their power and wealth, are able to find many social, political, and economic methods to either eliminate or severely curtail their exposure to risk.

Capitalists work hard at minimizing and socializing the risks involved with their business adventures. When things inevitably go wrong, they are usually shielded from the consequences. Here we see another example of the profound limits with the methodological individualism championed by the marginalists. Viewing people as a bunch of disconnected units with random tastes and preferences is the epitome of missing the forest for the trees. Human beings are aware and alert to the most important fact about our social existence: we have to share the world with other people and those people have a huge influence on what we do in life, whether we like it or not.

CHAPTER 3

Theories of Economic Growth and Development

Under capitalism, the doctrine that all economic growth is good represents an article of faith. Economies are supposed to grow so that more people can be absorbed into the labor force, so that corporations make higher profits, and so that the state receives expanding tax revenues. Growth is an imperative for capitalists, driven by their desire for ever-increasing capital accumulation. Assuming all economic agents are doing the right thing, economies will continue growing, but if there is a slowdown, prudent fiscal and monetary policies by governments can restore high growth rates. This is how the typical story goes.

I have argued that economic growth is a function of the aggregate energy flows that economic systems can generate. And to generate more work usually requires more energy conversion, unless the economy can produce substantial efficiency improvements. In general, growth fundamentally depends on the energy flows keeping an economy going. If those flows collapse or weaken, economic growth will follow the same path. When economies enter recessions and depressions, the defenders of capitalism grasp for a million different explanations. They have fantasized about markets as abstract and perfect entities, capable of all sorts of incredible wonders, like full employment, free trade, and endless profits. Meanwhile, they have neglected the underlying physical nature of economic systems, and in so doing they are utterly incapable of understanding how those systems work. With capitalism, economics has been transformed into a metaphysical philosophy, the goal of which is not to provide a scientific explanation for anything, but to create a propaganda machine designed to protect the wealth and power of the most elite groups in the world.

Any scientific explanation of economics must begin with the realization that the flows of matter and energy dictate the structure of all economies, not the invisible hand of the market. The economic features of the world are emergent properties shaped by underlying physical realities and ecological conditions, making an understanding of these conditions critical to any basic understanding of economics. In particular, we must understand that waste and dissipative losses can no longer be treated as mere "externalities" and the "costs of doing business," given the importance of these energy losses in shaping the dynamical evolution of economic systems.

The Neoclassical Theory of Growth

The intellectual foundations supporting the incessant drive for more economic growth derive from neoclassical economics. We have already examined the fundamental flaws of neoclassical theory from the perspective of its phony assumptions and its mathematical inconsistencies. In this chapter, however, we focus on the cardinal sin of this intellectual train wreck: its rejection of physics and its ignorance of the natural order. More than content to dabble in the falsehoods and contradictions of neoclassical theory, the economist Robert Solow developed one of the first major neoclassical models to describe how economic growth happens.[1] In these versions of neoclassical theory, the production inputs of capital and labor combine to produce outputs, or finished goods that are traded in the economy. Growth in capital leads to more output, but depreciation in capital assets also drags down a portion of that output. The economy eventually reaches a stationary state where growth and depreciation balance each other and there is no more growth. In order to produce continuous growth, neoclassical theory argues that the economy needs a steady stream of technological progress, defined as a gain in total productivity. This gain implies that productive output increases as a result of technological innovation, even while the productive inputs are held constant. Solow came up with a mathematical scheme for detecting the impact of this technological growth on changes in GDP. Although his work earned widespread acclaim from other neoclassical thinkers, many of his critical results can be safely ignored because they relied on useless "production" functions that explain absolutely nothing about production, considering how they're tested under empirical conditions that make them little different from an identity equation, the equivalent of saying that one equals one.[2] In addition, "real" GDP is not a reliable measure of economic growth, so one can never really know whether an economy has grown or shrunk in the context of neoclassical theory. The irony here is too powerful to pass up: a theory of economic growth that cannot actually tell you whether an economy is growing, or at the very least cannot determine its rate of growth.

Extensions of Solow's original work typically used the productive inputs of capital, labor, and technology to explain how growth happens. Energy and natural resources also entered this combustible mixture. Energy was either subsumed under the three main inputs or it was treated as a separate input in and of itself. To this very day, the marginalists regard the production inputs as being largely independent of one another, meaning that they can be substituted as necessary in order to maintain or to boost current levels of production. If societies are running short on natural resources, neoclassical theory states that these shortages can be overcome through technological innovation, efficiency gains, or other forms of substitution. Indeed, neoclassical economists tend to assume that the long-run sustainability of capitalism is materially possible, and all we need to do is figure out the social and institutional arrangements that would ensure that sustainability.[3]

It's this unshakeable faith in growth and human ingenuity that drives the climate-economy models of economists like William Nordhaus, who won the Nobel Prize for Economics in 2018. Nordhaus became famous for developing the "Dynamic Integrated Climate-Economy" model, abbreviated as DICE. Models like DICE are known as integrated assessment models (IAMs) and are used by the Intergovernmental Panel on Climate Change (IPCC) of the United Nations to make forecasts about the potential economic impacts of climate change. They thus have a huge impact on how politicians and countries worldwide assess the dangers associated with our age of ecological turbulence. Nordhaus measures the potential economic consequences of climate change through a "damage function" that purports to calculate the effects of rising temperatures on GDP.[4] The model is inherently useless because it's anchored to the flawed premise that GDP somehow measures the "aggregate scale" of an economy, when in reality it does no such thing. It's on the basis of this flawed reasoning, for example, that Nordhaus dismisses the dangers of climate change to global agriculture and forestry, since they "only" contribute around 4 percent to global GDP. Of course, if we run out of food, everyone starves to death. If we run out of Amazon and Walmart, we'd find a way to manage. Industry-level GDP shares are very often *not* reliable indicators of economic importance, and that's certainly the case if you're trying to understand the basis on which global civilization survives. It's precisely that 4 percent of global agriculture that makes the other 96 percent possible, so if the former disappears, so does the latter. Certain economic sectors and products act like critical gateways and bottlenecks for other economic activities, even if they themselves constitute a small share of total GDP. That's exactly what happened with the coronavirus vaccines, which reactivated large parts of the global economy as people felt safer to go out and enjoy their lives again.

For the moment, let's ignore the point about the limitations of GDP and focus on the other parts of the model. The damage function in the DICE model has a quadratic dependence on temperature change, but includes absolutely no factor to account for ecological tipping points, such as rapidly rising sea levels, thawing permafrost, or massive disruptions in ocean current oscillations. To pick an extreme example, if average global temperatures rise by an average of 10 degrees in 10 years, it would mean the extinction of the human species. The DICE model would instead treat this scenario as just another routine hit to GDP. And that's not an unfair exaggeration. Here is Nordhaus himself: "Including all factors, the final estimate is that the damages are 2.1 percent of global income at a 3 [degrees Celsius] warming, and 8.5 percent of income at a 6 [degrees Celsius] warming."[5] It's important to note that this 8.5 percent decline would be spread out over 100 years, so the actual projected annual decline in GDP growth would be less than 0.1 percent, a complete trifle. Considering how the Nobel Prize in economics has become an intellectual atrocity, it's not so surprising that Nordhaus managed to win it for an equation that a remedial algebra student could have scribbled down in detention. The most appalling fact in this sordid saga is that Nordhaus incorrectly cited a paper from a group of climate scientists to support his position that the damage function should not account for tipping points.[6] In reality, the climate scientists he cited very much supported the idea of ecological tipping points and were trying to resolve some uncertainties over how to calculate them in that very paper.[7] As the authors of the paper make painfully clear:

> Society may be lulled into a false sense of security by smooth projections of global change. Our synthesis of present knowledge suggests that a variety of tipping elements could reach their critical point within this century under Anthropogenic climate change. The greatest threats are tipping the Arctic sea-ice and Greenland ice sheet, and at least five other elements could

surprise us by exhibiting a nearby tipping point. This knowledge should influence climate policy.[8]

His models have many other problems. Nordhaus naïvely assumes that the vast majority of all economic activities won't be affected by global warming, which recent history has shown to be patently false.[9] There is absolutely no single area of human life that will be spared from the planet's accelerating thermodynamic instability. More important, Nordhaus has made a rather pathetic choice for the discount rate. As seen in the previous chapter, neoclassical theory assumes that money has some inherent and eternal time value, meaning that a dollar today is worth more than a dollar a year from now. Neoclassicals therefore apply a discount rate to determine a "present value" of the costs associated with expected ecological damages in the future. Discounting is supposed to be some kind of semi-objective way of relating the future and the present, but in reality, it's just a social ritual through which capitalists rationalize returns on their investments. Nordhaus uses a high discount rate of 6 percent on future ecological catastrophes not because he has any good evidence to back up that decision, but simply because using a high discount rate is a time-honored trick for producing smaller present financial values, making it appear as if rapidly rising global temperatures are not a serious threat that humanity should be concerned about. Nordhaus essentially assumes that the present value of future ecological costs is extremely low, thus justifying a laissez-faire attitude on what's actually a planetary emergency. Nordhaus has also criticized other economists for using smaller discount rates and therefore showing higher current social costs from future greenhouse gas emissions. In general, his understanding of climate science and ecological dynamics verges on the infantile. Thankfully, other prominent economists have reached different conclusions. In 2013, Robert Pindyck launched a devastating attack against the integrated assessment models used by the IPCC to forecast the consequences from global warming.[10] Pindyck correctly claimed that these models "are so deeply flawed" that they're basically "useless as tools for policy analysis." Fortunately, there's no need to play these silly pseudointellectual games, because the simple truth is that money has no time value. The economic decisions we make are not, for the most part, a reflection of personal time preferences; they are a reflection of class and power dynamics within the capitalist order. The way to know that global warming is a massive and potentially existential threat is not through vapid and contorted calculations of present values, but through rigorous studies in climate science and through concrete examples of societies worldwide being decimated as we speak. Humanity is staring down the barrel of a gun while the neoclassicals are fiddling with ridiculous models and useless abstractions.

The Limits of Substitution

Let's consider the central claims of neoclassical growth theory more closely. For a highly simplified toy model of how the theory works, consider your local pizza store. According to neoclassical theory, the pizza store can maintain or boost current levels of pizza production in the face of any shortfall. A shortage of workers can be overcome by adding more ovens. A shortage of cheese can be overcome through technical improvements that yield more efficient methods of making cheese. A shortage of electricity can be overcome by increasing labor productivity, perhaps by training the workers to make the pizzas faster under the new time constraints. Everything can be replaced. Everything can be substituted, seemingly without end. These ideas and principles represent fundamental assumptions in neoclassical economics, and they are often used to explain the relationship between energy conversion and economic growth. If there are no hard limits to substitution, then it would be possible for our economies to keep growing even in an ecosphere with declining quantities of natural resources and with highly chaotic, nonlinear ecological consequences that result from the enormous energy losses of capitalist societies. In other words, better technologies and higher efficiencies would always be available to boost production, regardless of any depletions or instabilities in the wider natural world caused by those productivity gains. To chip away at the elaborate fantasies of neoclassical growth theory, it helps to begin with some basic physics. We have already seen that the most fundamental limits to substitution come from thermodynamics. Thermodynamic limits impose constraints on the maximum efficiency of energy flows through technological systems. Car engines, power plants, and photovoltaic cells are all limited in their capacities to convert one type of energy into another. Technological progress cannot overcome these limits; no car engine, for example, can ever be more efficient than an engine running on the Carnot cycle.[11] Aggregate efficiencies for entire economies are highly inertial over time because improving them substantially requires enormous investments that would disrupt the reigning economic order.

Once a society has settled into a particular energy structure, changing it much further becomes a daunting task because of elite classes and groups that rely heavily on that structure for their wealth and influence. We can look to the recent experience of Germany for a prominent case study. In 2000, the German government launched its ambitious *Energiewende,* a comprehensive plan to reduce greenhouse gas emissions by shifting energy production toward renewable sources, such as wind and solar.[12] For a time, the program made some notable achievements. Compared to 1990, greenhouse gas emissions had declined 28 percent by 2017. That same year, renewables reached a 13 percent share of primary energy consumption. These numbers are impressive overall, but the progress that's been made recently is far less enviable. Germany only managed to reach its 2020 climate targets because of the global recession induced by the coronavirus pandemic. And once we dig into the numbers a bit deeper, even those that look impressive come with huge caveats. For example, a large portion of the reduction in carbon emissions since 1990 can be attributed to the collapse of heavy industry in East Germany after reunification.[13]

From 2010 to 2018, greenhouse gas emissions from Germany hardly changed. The short-term variabilities associated with wind and solar power have opened up problems related to electricity storage. Prices fluctuate dramatically depending on weather conditions. To compensate for these and other issues, Germany began sabotaging its energy program by constructing a series of new coal power plants when the coal industry pressured Merkel's government to relax its policies. The German example offers an important lesson: the necessary substitution of fossil fuels with renewables will never come fast enough under the market logic of capitalism.

Another major limit to substitution comes from the ecological instabilities associated with excessive levels of economic growth. These instabilities can combine to pump and amplify existing natural phenomena. Economies absorb energy from the natural world and then exploit that energy for cycles of production and consumption. For highly energy-intensive economies, these cycles necessarily yield extensive levels of waste and dissipation, or energy losses that are dumped back out to the environment. These energy losses are not useless. Under the right circumstances, they can power the formation of other natural dynamical systems, including everything from viruses and bacteria to wildfires and hurricanes.[14] These highly chaotic effects associated with energy-intensive economies are largely ignored and dismissed by neoclassical theory, even though they have often played a central role in the evolution of human history.[15] The convergence and aggregation of multiple natural feedback loops can produce what Marx called a "metabolic rift" between nature and society, which means that the ecological basis of civilization steadily erodes under profit-seeking and energy-intensive development that does not care about replacing what it extracts.[16] The natural world has major tipping points that we should not cross, but indefinite economic growth through substitution virtually guarantees that some of those critical thresholds will be breached, threatening the broader ecosphere that supports human civilization.[17] The overriding obsession with growth continuously forces neoclassical theory to marginalize and to dismiss the inevitable consequences that accompany energy-intensive economic expansion. These consequences include everything from air pollution to plastic trash in the oceans, and they provide another major warning sign for capitalism: even with the most brilliant fiscal and monetary policies, the ecological effects associated with our energy losses could still become powerful enough to disrupt the normal order of our economic systems. Of course, nature has a million ways of injecting chaos into our world, from epidemic diseases to wandering asteroids. But the critical point is that humans have far less control over the movement of a wandering asteroid or an exploding volcano than we do over decisions like where to live, what to eat, what car to drive, or whether to drive a car at all. We should focus on fixing our mistakes, not resigning ourselves to committing more mistakes just because other natural systems could destroy us in the future. The ultimate goal of civilization should be to create an ecological project rooted in equality and stability. Our aim should not be to destroy civilization itself, but to save it from capitalism.

Consider another problem. Substitution can occur quite frequently at small and restricted scales of economic activity. A pizza store can always substitute certain ingredients for others. A homeowner can substitute heating fuel for insulation. A company can replace older light bulbs for more efficient lighting in its offices. And even some countries can substitute various forms of wealth for others, at least temporarily. The island nation of Nauru, located in the Pacific, northeast of Australia in Micronesia, provides a classic example that highlights the central themes of the debate. In the twentieth century, Nauru possessed vast deposits of phosphate, which is highly prized as an agricultural fertilizer. These deposits were extensively mined, depleted, and then traded in global markets, allowing Nauru to reach one of the highest standards of living in the world by 1990.[18] Nauru converted a portion of its earnings from the phosphate trade into a public trust fund, which invested in manufactured capital through financial markets. However, its sky-high standard of living collapsed sharply after the phosphate had vanished, along with most of the money in the trust fund.[19] Nauru offers a cautionary tale for the world. If global civilization runs out of natural resources, we cannot replace them by investing in commodities through financial markets. People can't eat money. Substitution in the long run may be possible at the microlevel of economic activity, but long-term macrolevel substitution is downright wishful thinking.

We can also understand the limits to substitution on a global, macroscopic level by considering a specific example: growth in the context of a global economy meeting its electricity needs through the consumption of solar power. There are fundamental limits to the amount of solar energy absorbed by solar panels that can be converted into useful electrical energy. Most commercial photovoltaics convert roughly 15 percent of the solar energy they absorb into electricity; the remaining energy balance is lost as heat and infrared radiation.[20] The theoretical efficiency limits for the most advanced photovoltaic designs are just under 90 percent, a number that not even the latest laboratory experiments have come close to matching.[21] But suppose neoclassical theory is right about its eternal commitment to technological progress and that eventually we do manage to produce photovoltaics that are 90 percent efficient at converting solar energy. Once all theoretical efficiency limits are actually realized, boosting electricity production even further would require the construction of new solar panels, which takes up more land. As the Earth has a finite surface area, indefinite growth would not be possible even with the proliferation of renewables. This argument underscores the central point that renewable technologies are important, but they cannot succeed in solving the global ecological crisis under the economic regime of capitalism, which is completely reliant on the false promise of eternal growth in production and consumption. Substituting renewables for fossil fuels while pushing for more growth would still lead to the total ruin of global civilization in a few centuries.

Many neoclassical economists love to pretend that technological innovation can yield greater "qualitative growth" without any corresponding "quantitative growth."[22] On the basis of improving knowledge and technological growth, they believe that the monetary value of stuff can keep increasing even as the quantity of stuff itself remains stable. But what they fail to grasp is that technological innovations do not happen magically; they also require energetic conversions. You can't accelerate technological change without producing "more stuff" to feed into the process of innovation. Changes to the production cycle are dependent on the stock of electrical, chemical, and mechanical energy available for research and training. A coder sitting in front of a computer writing a new program needs energy to think and type. The computer itself needs electricity to continue operating. No possible improvement can be made to computer programs without a continuous stream of energetic conversions. Expansions in productivity require energy flows, meaning that all forms of technological change are intertwined with the energetic transformations that facilitate human existence. Technological changes are physically embedded in greater knowledge among people and the development of more productive assets, both of which need energy and material flows to continue operating. Thermodynamic limits also constrain the extent to which these flows can be reduced while sustaining labor and capital. In short, technological changes themselves are subject to hard physical limits, along with the "qualitative growth" that can be derived from them. Power plants provide one of the most well-known examples of the limits to technological growth. They have been hovering near their peak efficiency ratings for decades and getting them to go further has proven to be extremely difficult.[23] The failure of breeder reactors for nuclear power plants highlights another prominent technological bust, and plenty of other exotic technologies, like fusion reactors, will inevitably end up in the same category. The bloated profit margins of capitalism depend critically on the energy-intensive basis of its entire existence. Take away that basis and there is no more capitalism.

The Illusion of Decoupling

Economic growth cannot continue forever, but capitalists need to pretend that it can. Capitalism cannot acknowledge any natural limits to economic growth, for that would mean acknowledging its ultimate demise. To keep up the pretense that capitalism represents a quasi-eternal and invincible system, most political leaders and economists who support the current order have begun reciting a series of elaborate narratives about the relationship between human economies and the natural world. These narratives, bolstered by neoclassical economic theories all along, generally revolve around the central idea that we can *decouple* economic growth from the material needs of human civilization. It's the perfect message for and from the ruling classes: capitalism is sustainable, therefore nothing big needs to change. Until the late twentieth century, economists generally understood that more economic growth required the use of more energy and materials. But as the postwar compromises between labor and capital began collapsing in the 1970s and 1980s, the economic theories started to shift in emphasis and direction.

Inspired by neoclassical theories, a new generation of economists began to argue that economic growth could continue without requiring the consumption of additional resources from the environment.[24] They argued that we could reach this economic nirvana by doing more with less, by investing in clean energy, and by developing energy-efficient technologies. In short, they were arguing for nothing less than the long-term sustainability of capitalism, ignoring all of the science and evidence piling up along the way. At a basic level, pundits and economists generally define decoupling as a process in which the size of the economy keeps expanding while resource impacts, usually either carbon emissions or primary energy consumption, are declining. Common pieces of evidence cited by proponents of the decoupling narrative are charts over a short period of time showing carbon emissions falling while GDP is rising. Beyond emissions and energy consumption, some ecological economists have also looked at how GDP relates to the consumption of raw materials, although other ideas and concepts related to decoupling also circulate in the literature, reflecting the general ambiguity surrounding the issue. A distinction is often made in the literature between *relative* and *absolute* decoupling. Relative decoupling occurs when harmful impacts are rising at a slower rate than economic growth. By contrast, absolute decoupling occurs when harmful impacts are declining in absolute terms, even as the economy keeps expanding.

There are several major problems with the decoupling narrative. First and most important, GDP is not in any sense an accurate indicator for measuring "real output" and biophysical scale, and thus it's totally irrelevant that it may be decoupling from greenhouse gas emissions, pollution levels, or the average number of tooth fairies active on any given night. Because GDP does not and cannot say anything about the real domain of biophysical production and distribution, it would be false to conclude that economic growth has decoupled from harmful environmental impacts—since, again, changes in real GDP do not accurately measure changes in biophysical scale, and therefore a positive change in real GDP does not actually correspond to biophysical growth.

Second, even *if* GDP represented an accurate indicator of economic scale, temporary divergences between GDP and harmful environmental impacts do not indicate that decoupling is permanent. Consider the following fact: beginning in 2015, life expectancy in the United States experienced a steady decline that substantially worsened with the onset of the pandemic.[25] It has been the worst decline in American life expectancy in over a century. The United States now has a lower life expectancy than China. The American economy grew quite a bit during this period, if you believe the "real" GDP figures. But the press did not blare out the sirens declaring that life expectancy has "decoupled" from economic growth. Such an admission would raise an unthinkable prospect for the reigning plutocracy: the lives of common people might actually be getting worse while some billionaires are becoming even richer by selling the rest of us more stuff that does not improve our lives. However, when two or three years of mixed and uncertain data suggest that the rise in harmful global emissions has slowed down while economic growth has continued, the story gets blown out of proportion and becomes a powerful, if false, causal narrative about how capitalism can be ecologically sustainable. We should resist the temptation to make grand conclusions about the world when we notice marginal trends over just a few years.

Third, the datasets on global greenhouse gas emissions are so uncertain and unreliable that it's simply impossible to know whether there have been any genuine instances of absolute decoupling between GDP and harmful environmental impacts. Many popular media sources, like Our World in Data, show GDP growth for certain countries temporarily decoupling from *carbon dioxide* emissions and then claim victory for the mediocrities of late-stage capitalism.[26] But the recent acceleration in global warming that we've witnessed in this century has also been heavily driven by emissions of methane, nitrogen oxides, and synthetic gases, like fluorinated gases. Just because a country's GDP temporarily decoupled from carbon dioxide emissions does not mean that it also decoupled from *all* greenhouse gas emissions. This point cannot be emphasized enough. Some of the most prominent data sources regularly used by researchers, like the Global Carbon Project, don't even measure or estimate emissions of fluorinated and other synthetic gases, which are almost certainly very powerful drivers of global warming.[27] And even in cases where there are official estimates, the numbers are probably bogus. In the United States, the Environmental Protection Agency (EPA) measures fluorinated gas emissions based on self-reporting by corporations. (And if there's one thing we know about big corporations, it's that they never lie about anything!) Corporate accounts on energy and pollution figures are mostly useless because large corporations usually ignore the upstream and downstream effects of their operations. They may also deliberately obfuscate or undercount their negative environmental impacts as a way of avoiding regulatory scrutiny, like the automaker Volkswagen did when it developed software designed to cheat on governments tests measuring tailpipe emissions of harmful gases.[28]

Take the case of methane as a concrete and important example. It leaks routinely through oil and gas wells, pipelines, and other fossil fuel infrastructure. From 2012 to 2018, the Environmental Defense Fund conducted a huge study that attempted to track actual methane emissions in the United States. The study found that the oil and gas industry emitted at least 60 percent more methane than what they had reported to the EPA.[29] And lest you think this is a fluke, the International Energy Agency released a damning report in 2022 on methane emissions showing pretty much the same thing. Using satellite data, the IEA found that methane emissions from the energy sector are 70 percent higher than what's officially reported by corporations and national governments.[30] Methane and fluorinated gases have a much higher "global warming potential" (GWP) than carbon dioxide, which means that even though they may be emitted in far smaller quantities than carbon dioxide, they can still trap vastly more heat in the atmosphere in the short run. Indeed, many climate scientists and researchers now believe that the large atmospheric concentrations of these poorly tracked gases are responsible for a huge chunk of the recent global warming patterns. The doubling of methane in the past two hundred years is probably responsible for 20 to 30 percent of the observed global warming since the Industrial Revolution.[31]

The manipulation of GHG emissions data is a fundamental and endemic problem plaguing virtually every country and virtually all government databases. A major investigation conducted by reporters at the *Washington Post* in 2021 revealed that the world's countries are, on an annual basis, collectively emitting anywhere from 8.5 billion to 13.3 billion tons of carbon dioxide *more* than what they've officially reported.[32] The level of deception can sometimes border on the comical. Malaysia, for example, claimed that its 68,000 square miles of forests provided an annual carbon sink of roughly 240 million tons of carbon dioxide, roughly the same amount as claimed by Indonesia, which has a forested area five times larger. The world's leaders have become experts at holding glamorous meetings in fancy hotels and conference halls, pledging to reduce emissions in front of the cameras while at the same time greenlighting their corporate donors to inflict maximum damage on the planet. Average global temperatures are now on a path to beat the critical threshold of 1.5 degrees Celsius above pre-industrial levels by the 2030s.

Western corporations are particularly notorious for making grandiose energy and emissions claims that rely on highly deceptive and manipulative accounting tricks. In its 2015 *Environmental Sustainability Report*, Apple claimed that "all our data centers [since 2012] have been powered by 100 percent renewable energy sources. That means no matter how much data they handle, there is a zero-greenhouse gas impact on the environment from their energy use."[33] This claim was completely false. Take the case of Apple's data center in Maiden, North Carolina. Apple purchased the electricity used to operate this data center directly from Duke Energy, the main power supplier in North Carolina. But Duke's grid is largely a mix of nuclear and coal power plants, with some solar and wind farms also making a small contribution. Whenever the Maiden data center was operating, carbon emissions were being generated. Apple argued that it managed to "offset" its fossil fuel reliance at Maiden by producing some electricity through a solar farm and fuel cell system located on-site, but this argument is a silly red herring because it distracts from their central claim that operating the data center had no emissions impact on the atmosphere. It most definitely did. What's worse, at the time of the 2015 report, Apple's precious fuel cell system at Maiden was actually powered by *natural gas* from the company Piedmont Natural Gas.[34] Apple is not an isolated example. Amazon's massive data centers in Northern Virginia also rely heavily on electricity purchased from Dominion Power, which unsurprisingly uses fossil fuels for much of its power output.[35] The simple truth is that major corporations are de facto addicted to fossil fuels, regardless of what they claim. Big corporations have adopted many gimmicks to pretend that they're doing something about the environment. For example, big banks love to brag about how they'll reach net zero emissions by 2050, a meaningless promise that's become a capitalist catechism.[36] Their actions don't match their rhetoric. In the six years since the Paris Agreement was signed, the world's sixty largest private banks provided a stupendous $4.6 trillion in financing to fossil fuel companies, with $742 billion worth of loans and investments in 2021 alone.[37] This money paid for things like exploration costs, drilling, fracking, refining, storage, and just about every other conceivable operation that keeps the fossil fuel industry going. In recent times, the rise of "carbon offsets" has become another favored corporate greenwashing strategy. In this scheme, companies supposedly "offset" their economic activities that release greenhouse gases by paying money to private organizations to invest in various environmental projects, like protecting an endangered forest or building a new wind farm. But the biggest flaw of this entire strategy is that companies often pay for projects that would have happened anyway, so they don't actually remove any additional carbon dioxide from the atmosphere. In 2020, banking giant JPMorgan claimed to have reached net zero emissions, in large part because it had "bought $1 million worth of offsets to protect a forested area in Pennsylvania known as the Hawk Mountain Preserve."[38] But as its name clearly suggests, the area was never in any danger of deforestation, so paying to protect an area *that was already protected* did not compensate for any of JPMorgan's emissions. Study after study has shown by now that carbon offsets and credits are an abysmal failure.[39] They've simply become a convenient license that corporations can purchase at will while continuing to trash the planet. The only effective way to reduce emissions is to downscale and restrain the activities that are producing those emissions in the first place. When it comes to corporations and their emissions accounting, our motto should be: *Never trust, always verify*. Corporate greenwashing is a trend that's here to stay, and corrupt governments will allow the corporate world to get away with planetary ecocide unless the public demands an alternative path.

The rise of palm oil plantations in Indonesia is another instructive example of corporate and government manipulation when it comes to ecological accounting. In the early twenty-first century, production of palm oil boomed in Indonesia, destroying much of the country's pristine tropical rainforests along the way because of increased demand from Asian countries and legislative changes in the United States, which called for reducing the fossil fuel content in gasoline and replacing it with biofuels, like corn and soy.[40] Politicians and many fuel "experts" argued that the emissions would net out, since plants absorb carbon dioxide from the atmosphere as they grow, and then this carbon dioxide is released through tailpipe emissions once the fuel is burned in vehicles. The reality of global land use trashes that simple story. Crops are not magically grown on land that has never been touched before; forests, bushes, and other plants have to be cleared in order to make room for intensive agriculture, and that destructive process releases vast quantities of carbon dioxide. Furthermore, new plantations are typically established on top of older peatland soils, and disturbing these soils can also release large amounts of carbon dioxide. Using biofuels only exacerbates GHG emissions. Yet many politicians in the United States continue to believe that burning biofuels is a good idea. Their misguided opinions certainly have nothing to do with the huge financial lobbying and political pressure that they've received from agribusiness, which wants to mandate the use of more biofuels as a way of increasing sales and profits.

The decoupling narrative is a distracting fantasy from the hard reality that the only plausible path to avoid an impending ecological catastrophe is through a major transformation of our political and economic regimes. It's a distraction that exists in many forms, especially those that claim that economic growth can be decoupled from energy use. Unsurprisingly, these narratives are also false. All economic activities require energy. Neoclassical thinkers make a category mistake when they view energy as just another input of production, as opposed to seeing energy for what it is: the fundamental basis of all economic activities, including production. Energy is what makes all methods of production possible in the first place. In this fundamental sense, economic activities cannot be decoupled from energy use, for that would be like asking economics to step completely outside the laws of physics, which is a clear absurdity. But this clear absurdity is what neoclassical theory implies can actually happen: by artificially detaching capital and labor from energetic constraints, it effectively severs any and all links between physics and economics. For example, when economists and media outlets show plots of GDP growth diverging from energy conversion, they are really showing GDP growth diverging from *primary energy consumption*. They then assume that this alone somehow proves that economic growth has become detached from energy use. Of course, we've already seen that real GDP is a useless aggregate quantity that measures neither economic growth nor scale, but let's set aside this point for a moment and indulge the other side. A divergence between changes in GDP and changes in primary energy consumption does not mean that economic growth has decoupled from energy use, simply because primary energy consumption ignores the secondary and tertiary energy flows and conversions that sustain all economic activities, such as watching a movie on your television, browsing the internet on your computer, or driving your car to the grocery store. Thinking exclusively in terms of primary consumption makes it seem like energy simply disappears after its initial entry point at some power plant. Capital, labor, and technology then magically take over the process of production, operating on fairy dust the whole way through. The reality is that the activities of capital, labor, and all forms of technology are dependent on energetic conversions, and this is the fundamental sense in which it's silly to claim that economic growth can be decoupled from energy use.

Let's look at the economy of the United States more closely to understand what's going on. In recent decades, American economic growth measured in GDP has continued, albeit at a declining rate, even though per capita primary energy consumption has declined.[41] In addition, costs associated with primary energy consumption increasingly represent a declining share of American aggregate demand, again measured in GDP. From these observations, many economists and pundits have concluded that energy use and economic growth have decoupled from one another.[42] But even a quick analysis of the underlying energy shifts in the American economy reveals the falsehood of this narrative. An economy that starts using natural resources with higher energy efficiencies and larger power densities can experience growth even as primary energy consumption declines. Understanding this process would be difficult, perhaps even impossible, if we only looked at primary consumption, which totally ignores conversions. But once we consider that burning a smaller quantity of natural gas, for example, can still yield more electricity than burning a larger amount of coal, then the significance of conversions becomes immediately apparent.

Resources and devices with larger power densities can convert more useful energy for economic activities, some of which constitute the basic elements measured by GDP. Economists like David Stern, Robert Ayres, and Robert Kaufmann, among others, have shown that growth in American energy conversion is tightly coupled with growth in aggregate demand once differences in energy quality are factored into the analysis.[43] The energy crisis of the 1970s motivated the United States to reduce per capita oil consumption and to focus on efficiency gains by using other natural resources. These efforts resulted in a trajectory of increased natural gas consumption, which is much cleaner and more efficient as an energy source than coal. The switch to natural gas and the increasing proliferation of renewables both helped to substantially reduce carbon emissions. After arguably peaking in 2005, greenhouse gas emissions in the United States had fallen 14 percent by 2016.[44] But the declines gradually stalled, and emissions in 2018 actually increased by more than 3 percent, the largest rise in eight years.[45] A hyperactive transportation sector, always critical to economic growth, was the leading culprit behind that particular surge. A temporary dip in GHG emissions during the pandemic was followed by huge increases over the 2021–2022 period, for both the United States and the world at large.[46] Recent American experience further reinforces the notion that large-scale reductions in emissions are virtually impossible under an economic system that prioritizes growth above anything else. The intense sociopolitical pressure to increase consumption and production can lead to rising emissions even in the context of macrolevel efficiency gains and technological innovations. Figure 3.1 shows this basic problem on a global scale, as renewables only constituted roughly 11 percent of the world's primary energy consumption in 2019 despite decades of rapid growth.

For the world as a whole, a strong positive relationship exists between primary energy consumption and economic growth, even with the underlying flaws in the way we measure that growth. Numerous studies for various countries and regions indicate that this relationship is fundamentally causal.[47] Over the last few decades, the rate of global economic growth has started to slow down, mirroring the declining rate of growth in global energy conversion. Some major economies, like those of Japan and the European Union, have already entered periods of stagnation associated with very low growth rates and aging populations. Because these economies are currently dominated by corrupt financial sectors, they are generating uneven growth patterns that mostly enrich wealthy capitalists. By contrast, ordinary people are increasingly drowning in debt so they can finance the cycles and crises of capitalism.[48] Economic progress for the vast majority of society has come to a screeching halt.[49] The global economy may continue to grow at modest rates for the rest of this century, but the signs are already evident that our potential for future growth is limited and constrained by what kinds of energy sources we can collect from the natural world, as well as by the economic irrationalities of today's financialized capitalism. Capitalism is running out of steam, but not quickly enough to substantially reduce aggregate

FIGURE 3.1: Global Primary Energy Consumption by Fuel Source.

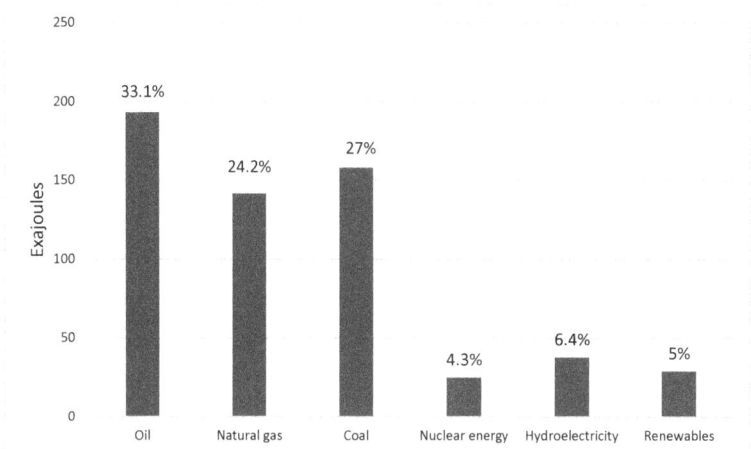

Note: An exajoule equals 1018 joules.
Source: BP, *The 2020 Statistical Review of World Energy*, p. 4.

emissions.

For a certain period of time, optimism on global warming was running high. Greenhouse gas emissions flatlined for several years and the upper echelons of the global economy started to believe that economic growth could actually be decoupled from harmful emissions. In 2016, the International Energy Agency triumphantly declared: "Decoupling of global emissions and economic growth confirmed."[50] What a difference a couple of years can make. In 2017, greenhouse gas emissions worldwide saw a sharp spike.[51] They rose again for 2018, at a faster pace than the previous year, all in the background of increasingly alarming scientific reports about the dangers of global warming.[52] Even some advanced economies that had supposedly decoupled growth from pollution witnessed higher carbon emissions in 2018. The year 2019 was hardly any different. It was only the arrival of a global pandemic in 2020 that momentarily broke these horrible trends, but the sudden shock has now worn off and the global economy is back to its energy-ravenous ways. Detaching emissions from economic growth has turned out to be a vastly more complicated problem than global elites originally believed.

A persistent albatross on this issue is the way that most elites talk about carbon emissions. When governments and some organizations measure greenhouse gas emissions, they often do so at the point of manufacture and production. If an American company sets up a factory in India to produce commodities that are then sold to American consumers, the emissions coming from that factory are credited to India, not the United States. This basic process of "geographic substitution," where corporations from the capitalist core transfer ecologically destructive manufacturing to developing nations with large pools of cheap labor, has been an important source, though not the only one, of the observed divergence between carbon emissions and economic growth in the Western world.[53] Studies that measure emissions at the point of consumption show virtually no decoupling whatsoever at various geographical scales of analysis.[54] For many countries, there can be enormous differences in per capita GHG emissions depending on whether the emissions are tracked at the point of production or the point of consumption. Consumption-based indicators, for example, show that Western and developed countries have much larger per capita emissions than production-based indicators. In any case, multinational corporations can only keep shifting production around so much before they run out of places to go. There are limits to geographic substitution as well.

The Cultural School

Neoclassical economic theories are not the only game in town when it comes to explaining the origins and dynamics of economic growth and development. Various scholars have invoked all kinds of theories to explain the differential growth trajectories of human societies, including the behavioral impact of cultural beliefs, the incentives provided by political and economic institutions, and the role played by natural factors like ecology and geography. Since this last one will be covered extensively in the

THEORIES OF ECONOMIC GROWTH AND DEVELOPMENT 127

next few chapters of this book, I'll focus on addressing the first two for now, beginning with cultural explanations. The historian Joel Mokyr argued in his 2016 book, *The Culture of Growth*, that the Industrial Revolution was largely a product of unique cultural and institutional forces acting in certain parts of Europe.[55] He held that the beliefs, values, and preferences of early modern Europe led to profound changes in behavior that hastened scientific advances and pioneering inventions. He also claimed that political fragmentation in Europe led to a marketplace of ideas that fostered creativity. Mokyr belongs to the "cultural school" of explanations for the development of the Industrial Revolution and modern capitalism. The most famous proponent of this school was the sociologist and historian Max Weber, who had argued that capitalism emerged in early modern Europe because Protestants had a strong work ethic geared toward boosting productivity. However, Mokyr's thesis is the most advanced version of anything ever proposed by the cultural school, and for that reason I'll focus on his work and its implications in this section.

Mokyr wants to resolve what he considers to be a puzzle in modern economic thinking. On the one hand, he acknowledges the importance of institutions in fostering economic growth and stability. Institutions could be things like law and order, private property rights, efficient governance, and contract enforcement. However, Mokyr argues that the Industrial Revolution, which started the era of modern growth, was primarily about technological innovation, and not just better resource allocation and efficient markets. He argues that institutional change, in the sense of gains from trade and specialization, cannot explain the full extent of modern economic growth. Before 1750, Mokyr argues, economic growth was largely "Smithian" in the sense that it heavily depended on the right mixture of traditional institutions. Adam Smith had argued that the government should protect private property and stop sabotaging the "invisible hand" of the market. But after 1750, Mokyr maintains that growth increasingly relied on innovation and "useful knowledge." Mokyr calls these growth dynamics "Schumpeterian," after the Austrian economist Joseph Schumpeter.

To explain these supposedly different growth patterns, Mokyr decided to look at the role of European culture in generating better scientific ideas and more technological innovation. Mokyr defines culture as a "set of beliefs, values, and preferences, capable of affecting behavior, that are socially (not genetically) transmitted and that are shared by some subset of history."[56] Mokyr describes the three main elements of culture as follows:

> First, beliefs contain statements of a positive (factual) nature that pertain to the state of the world, including the physical and metaphysical environments and social relations. Second, values pertain to normative statements about society and social relations (often thought of as ethics and ideology), whereas preferences are normative statements about individual matters such as consumption and personal affairs.[57]

In his book, Mokyr focuses on cultural attitudes and beliefs about our surrounding physical environment and humanity's relationship to the natural world. He also emphasizes that he's talking about the beliefs and preferences of a small group of elite intellectuals, whom he credits with making most of the major breakthroughs that would precipitate the Industrial Revolution. Mokyr then introduces a metaphor: the "market of ideas." According to him, intellectual leaders and innovators try to persuade "buyers" about the utility and superiority of their ideas, almost as if this was a market with goods and services. As these intellectuals persuade more people, their social reputation increases and their ideas become more dominant. Mokyr understands that he's not describing a real market, since there are no prices involved, but nevertheless maintains that his metaphor is useful because it allows for the discussion of concepts like competition and barriers to entry. Mokyr claims that any market of ideas can face bad incentives that hamper its growth. For example, new ideas can weaken the institutional authority and material prosperity of the ruling classes. That could lead to the ideas being branded as "heretical" or "dangerous." They might even be formally banned, and their proponents could be actively persecuted. Mokyr's central claim in the book is that, between 1500 and 1700, the European educated elite developed a cultural and institutional framework that allowed for greater intellectual innovation and the proliferation of more useful knowledge than before.

To explain how all this happened, Mokyr brings out his theoretical arsenal. The central feature of his theory is the idea of choice-based cultural evolution. Borrowing from evolutionary theory, Mokyr argues that ideas and other cultural units experience selection pressure from individuals who are making choices, almost as if they were picking items from a menu. People have certain biases that they acquire over the course of their lives. As they interact with others in society, they use a variety of methods to spread or defend their ideas. They may introduce risky proposals, try their hand at persuasion, engage in imitation, succumb to conformity, or show obedience to authority. Sometimes they can do many of these things at the same time. From 1500 to 1700, Mokyr says that several cultural variants arrived on the scene, such as Protestantism, heliocentrism, Cartesian dualism, and Galilean mechanics, among others. Some central tenets of the Enlightenment also emerged during this period. These included the belief in the possibility and desirability of human progress, the belief in the superiority of modern thinking over ancient and classical doctrines, and the belief that "useful" knowledge should be used for production, which established a new scientific agenda for applied research. Eventually, the market of ideas led to new cognitive equilibria and "meta-ideas," such as the importance of collecting facts and data, of looking for empirical regularities, of using mathematics and quantification for modeling phenomena, and of divorcing science from metaphysics and from other speculative philosophies. Mokyr believes that these cultural changes laid the groundwork for the Industrial Revolution. He argues that the European market of ideas changes in many ways from 1500 to 1700. It encouraged innovation through better incentives, became more competitive, and lowered entry barriers. But why did this market of ideas succeed?

To answer this question, Mokyr supposes that markets work effectively when they have an institutional foundation, which should specify the incentives for participants and the rules by which the market operates. The institution needs to both increase positive incentives, by finding ways of rewarding intellectual innovators, and to decrease negative incentives, by weakening the forces that would suppress innovation. What was this magical institution that allowed Europeans to leapfrog the rest of the world in economic performance and strategic power? "The Republic of Letters," as Mokyr calls it. Europe was not the first place in the world to have a market for ideas, but it was the first to find an ingenious institutional framework that spurred further growth and innovation, according to Mokyr. The Republic of Letters (RoL) was a transnational community of scholars and elite intellectuals who gradually established a vast communication network in which they exchanged letters and ideas. This elite group of thinkers included everyone from scientists and philosophers to theologians and alchemists. Sometimes the distinctions were blurry, but the point is that these people were highly educated and subscribed to the same cultural outlook. Mokyr calls the RoL an efficient institution that was not designed to be that way. In that sense, it functioned like an emergent system, a complex phenomenon resulting from much simpler interactions.

Mokyr calls the RoL a "weak ties" network in which people didn't know each other very well and therefore had to back up their claims with rigorous and persuasive evidence. According to him, the RoL succeeded because it was an open community in which many people could participate, it was a transnational community that did not belong to a single country, and it contested all knowledge claims through reason and experiments. Mokyr argues that the RoL was successful because it remained largely independent from political and religious institutions that would otherwise have harmed its progress. And it protected this independence thanks to the political fragmentation of the European continent. If a member of the RoL got in trouble somewhere, he usually had the option of moving to another country that wanted his services, playing off rival powers against each other in the process. Other critical components for its success included the printing press, the rise of an urban middle class, the growth of postal services, the European voyages and expeditions around the world, and competitive patronage among rulers who wanted the best officials. The competitive patronage system also involved universities and other institutions; everyone wanted to have the best and the brightest. This competitive environment held the prospect not just of gaining more reputation, but also of potentially securing a lucrative salary. In addition, Mokyr claims that the RoL created economies of scale: it grew into a pan-European institution that could rapidly disseminate knowledge across many countries at relatively little cost.

Mokyr occupies the "mushy center" with his core positions. He opposes what he considers to be the extremes of materialism, with its apparent emphasis on ecological and economic forces that supposedly leave no room for human agency, as well as the extremes of idealism. To his credit, he never enters the realm of idealist platitudes, with claims like "ideas drive history" or other such ridiculous notions. He understands that all ideas are embedded in a certain economic, institutional, and historical context. But that's about the best one can say for his overall argument. One of the biggest problems with Mokyr's thesis is that, even assuming it's largely correct, it does not explain why the Industrial Revolution specifically started in Britain. This is a problem that the author himself has acknowledged in the past. The failure highlights the profound limitations of using culture as a causal explanation. Any theory about the rise of capitalism and the emergence of the Industrial Revolution must explain the central role that Britain played in the entire process. Plenty of European societies had intellectuals exchanging ideas during this time; the challenge for the cultural school is to explain why this process led to the Industrial Revolution in Britain, and not in Portugal or Italy. But any serious attempt to formulate such an explanation will require some kind of materialism, like acknowledging the highly unusual economic and ecological circumstances that Britain faced in the early modern period. This implies that cultural explanations of the Industrial Revolution will always be provisional, incomplete, and overly focused on ideas and values that have no proper causal grounding.

Those who insist on prioritizing cultural explanations for the evolution of human history generally run into problems with causality. Let's explore some of these issues in the context of Mokyr's book. First, there's the demarcation problem. Mokyr's definition of culture is so broad and so vague that it could apply to virtually all conceivable aspects of human society, such as politics, economics, or philosophy. Consider the belief that everything in the world is made up of tiny particles called atoms, a belief that many Enlightenment thinkers held after the rediscovery of lost Epicurean writings. It's an important belief because it affected the work of several major scientists, including Isaac Newton and the chemist Robert Boyle. But is this belief a cultural or a philosophical belief? Maybe it's both. Philosophy could certainly be a part of culture. But it's easy to see how this can quickly escalate. Virtually all beliefs that people have overlap with culture in some way or another. The belief that financial profit is good for society can be seen as both economic and cultural. The belief that democracy is the best form of government can be seen as both political and cultural. What are the causal boundaries of culture? Where does it end and everything else begin? Mokyr does not address these questions, but in fairness to him neither do most people. The result is that culture acts like a black box capable of accommodating any and all preconceptions of the world. It becomes a very broad category that tries to say everything, but really says nothing.

A related complication for the cultural school is the problem of original ideas. Philosophers sometimes use the term *ideogenesis* when talking about the creation of new ideas. A major flaw in Mokyr's definition of culture is that it excludes all acts of ideogenesis, even though these mental activities can clearly affect an individual's behavior. Suppose someone comes to believe that giant ants live on the surface of Pluto. Let's assume that this person didn't get the idea from anyone else; she thought of it herself, and eventually she starts to believe that this idea is true, that there are real giant ants living on Pluto. She starts writing articles about it and convinces some other people that it's true, meaning that her belief is affecting her behavior and is also being socially transmitted. What we have here is a case where all subsequent acts of social transmission and behavioral change can be traced back to a single instance of ideogenesis, which is itself not a feature of human culture, per Mokyr's definition. But virtually all socially transmitted beliefs can be causally regressed backwards to some baptismal act of ideogenesis, which implies that nearly all cultural beliefs are derived from non-cultural beliefs, suggesting that culture is not the fundamental explanation of social and economic outcomes. Mokyr could conceivably get around this problem by simply modifying his original definition to indicate that beliefs can affect behaviors, but *they can also affect other beliefs*. This strategy won't work either, however. That's because people may have beliefs that affect behaviors, and by extension the behaviors of those around them, but without any corresponding social transmission. In other words, not all ideas and beliefs are meant to be shared. People form plenty of beliefs that they prefer to keep to themselves. Imagine a common situation: a boss who holds negative beliefs about his workers, but who doesn't want to share all of those beliefs for fear of creating a difficult and complicated work environment. The beliefs of this boss are still important because they may affect how he does his job or how he interacts with his employees, which in turn could affect their prospects for promotion and other aspects of their future careers. Restricted acts of ideogenesis are therefore critical in determining social and economic outcomes, especially when they involve powerful people, and yet these compartmentalized beliefs are excluded from Mokyr's definition.

The second problem for the cultural school is its implicit assumption that materialist and cultural explanations are at odds with each other. This position brings up several ontological questions: What are the fundamental constituents of culture? Are they ideas, material things, or something else? And what is the material world itself? Does culture exist within or outside the material world? To ask these questions is to wonder, in effect, if human beings are part of the natural world. If we are, then culture must be too. If not, then culture does not exist within the material world, and is therefore not constrained by it. It's reasonable to hold that human beings are part of the natural world because the laws of physics, as far as we know, apply to all of us. This fact further implies that we are complex physical systems, otherwise we could not be affected by physical interactions, like gravitational and electromagnetic forces. It also implies that our minds are complex physical systems, because all minds are shaped and constrained by interactions in the natural world. Indeed, this is the de facto position in modern neuroscience: that the human mind is a complicated *physical* system. By extension, then, our social interactions must be part of the natural world as well. Whenever we sleep, laugh, exchange ideas, or hike in the wilderness, we are doing so as natural creatures. Because culture is subsumed under the material world, we can examine it through causal analysis. We can ask lots of interesting questions about it. What caused the rise of that idea? What caused a particular social practice? Why did that religion become popular? Answering these questions will be extremely challenging most of the time. We do not have access to the chaotic neural dynamics that encoded the thoughts and memories of people in the past. But we can still sketch out a general solution to most interesting problems. For example, Polynesian societies had a rich mythology involving water and the ocean. Does anyone seriously think that this profound aspect of their culture can be decoupled from the fact that they lived on islands surrounded by the biggest ocean in the world? The lesson is simple: culture is bounded and constrained by external forces.

This brings us to the third problem. Because culture is bounded and constrained by our wider material existence, it can never serve as the ultimate causal explanation for anything. Materialists have no problem admitting that culture is important and that it shapes our lives in many ways; the Marxist philosopher Antonio Gramsci famously developed the notion of *cultural hegemony* to describe how the ruling classes under capitalism perpetuate their power by controlling and influencing the cultural levers that shape how people think. But materialists do not pretend that the chain of explanation simply stops right there. The cultural camp might contend, as Mokyr does, that culture is a high-level emergent phenomenon. It cannot be reduced to its underlying material interactions. But even if we accept this point, it does not show what the culturalists think it does, because every emergent phenomenon must emerge *from something*. Even if the whole has different properties from its parts, there would not be a whole if there weren't any interacting parts. All emergent phenomena require unique physical substrates. The elements and conditions that would produce a superfluid are different from those that would produce a superconductor. The implication for our argument is clear: culture can be invoked as a partial and provisional answer, but never as the ultimate solution to any profound question about the structure of society. All elements of culture must ultimately have non-cultural causal antecedents, otherwise cultural explanations would be stuck in an infinite regress. Not every aspect of culture can be explained by invoking another aspect of culture. There has to be something outside culture that breaks the chain.

To really drive home the point about why Mokyr's thesis is fundamentally wrong, consider another society with very similar conditions to the one he examined for his book: Ancient Greece during its peak phase in the Iron Age (c. 700 BCE to c. 300 BCE). The ancient Greeks lived in a politically fragmented world of competing city-states and wealthy colonies. Their elite intellectuals regularly exchanged letters, ideas, and even held public debates. Technological innovations in one part of the Greek world rapidly made their way to another; improvements and modifications usually followed the original designs. Consider that the Greeks invented the crane, the shower, the watermill, the world's first analog computer, known as the *Antikythera mechanism*, and even rudimentary versions of the cannon and the steam engine, both of which, unfortunately for them, had limited applicability in their original forms. Following the Bronze Age collapse, the Greek world enjoyed impressive economic growth rates for several centuries, bolstered by internal competition and an expanding trade network that carried Greek ships all the way to Britain. And yet, for all its incredible achievements, Ancient Greece was not the place where the Industrial Revolution happened. Why not? Because culture isn't everything. People can aspire to achieve anything they want. They may have grand ambitions and desires for more knowledge, wealth, and power. But if they don't have the necessary economic and natural resources to overcome the barriers that keep their ambitions in check, they are not going to produce the energy-intensive societies of the modern capitalist world. Greco-Roman civilization failed to fully industrialize because the necessary material and ecological conditions weren't there. Culture does play a crucial role in shaping economic performance, but only a secondary and provisional one. Ecological dynamics form the dominant causal matrix for all economic systems.

The Institutional School

Another prominent school of thought argues that economic development chiefly comes down to having the right rules and institutions for society. There are intellectual variations among institutionalists just like in any school of thought, but they generally prefer to see economic outcomes as the result of setting the right rules and incentives for people to follow. To institutionalists, societies succeed when they implement features like the protection of private property, market rules that foster competition, checks and balances in government, independent judicial systems, competitive multi-party elections, and so on. Two of the more prominent thinkers in the institutionalist camp have been the economists Daron Acemoglu and James Robinson. In their 2012 book *Why Nations Fail* Acemoglu and Robinson claim that "countries differ in their economic success because of their different institutions, the rules influencing how the economy works, and the incentives that motivate people."[58] According to them, "Inclusive economic institutions create inclusive markets, which not only give people freedom to pursue the vocations in life that best suit their talents but also provide a level playing field that gives them the opportunity to do so. Those who have good ideas will be able to start businesses, workers will tend to go to activities where their productivity is greater, and less efficient firms can be replaced by more efficient ones."[59] Acemoglu and Robinson are persuaded by the observation of the Austrian economist Joseph Schumpeter that *creative destruction* is a major driving force in modern capitalism, meaning that new ideas, firms, products, and technologies replace older ones that can no longer compete. Incidentally, Schumpeter actually admits to borrowing the idea from Karl Marx and Friedrich Engels, who argued in *The Communist Manifesto* that the recurring commercial crises of capitalism required the periodic destruction of older methods of industry and production to pave the way for newer methods and social relations in the next economic order.[60] To their credit, Acemoglu and Robinson don't just focus on economics, but also examine politics as central to the nature of social institutions, holding that "politics surrounds institutions for the simple reason that while inclusive institutions may be good for the economic prosperity of a nation, some people or groups, such as the elite of the Communist Party of North Korea or the sugar planters of colonial Barbados, will be much better off by setting up institutions that are extractive. When there is conflict over institutions, what happens depends on which people or group wins out in the game of politics, who can get more support, obtain additional resources, and form more effective alliances. In short, who wins depends on the distribution of political power in society."[61]

Acemoglu and Robinson do acknowledge the importance of what they call "critical junctures" in history, such as wars and pandemics. However, they only see these critical junctures as important because of their ability to amplify preexisting small differences among institutional structures in different countries, *not* because these critical junctures can fundamentally affect the process of institutional formation and destruction in the first place. Their fundamental unit of explanation is and always remains the institution. According to Acemoglu and Robinson, societies can succeed by establishing inclusive economic and political institutions that encourage prosperity across the social spectrum. On the flipside, societies start failing when corrupt elites establish extractive institutions that siphon off wealth from the majority of the population and redirect it toward themselves. This simple duality is the backdrop for their historical analysis, and they go through many historical examples in the book, from medieval Venice and Spanish America to England and North Korea.

Their most important case study is the economic evolution of England from the Middle Ages to the Industrial Revolution. They hold that England's economic rise boils down to the country supposedly developing inclusive political and economic constitutions that encouraged economic growth, market competition, and technological innovation. They cite various examples throughout English history, such as the signing of the Magna Carta in 1215 and the Glorious Revolution in 1688, for taking England toward a direction of political pluralism and away from a direction of political dictatorship. Over time, they argue, these inclusive and pluralistic institutions culminated in the Industrial Revolution, the ascendance of capitalism, and the birth of the modern world. As Acemoglu and Robinson sum up their views on the Industrial Revolution in England:

> The Industrial Revolution started and made its biggest strides

in England because of her uniquely inclusive economic institutions. These in turn were built on foundations laid by the inclusive political institutions brought about by the Glorious Revolution. It was the Glorious Revolution that strengthened and rationalized property rights, improved financial markets, undermined state-sanctioned monopolies in foreign trade, and removed the barriers to the expansion of industry.[62]

Institutional explanations of human history deserve more credit than their cultural counterparts for being specific, focused, and less prone to subjective judgments. Responsive institutions and governments are obviously important for economic development. No one wants to live under a corrupt government. However, most institutional theories still suffer from some of the same problems as cultural theories, especially regarding causality, emergence, naïve reductionism, and boundary problems. The simple truth is that institutions are merely a proximate cause of economic growth, not a fundamental one. They are dependent on a broader causal matrix of other material factors, many of which people don't control at all. Acemoglu and Robinson have no concrete explanation for what causes responsive or extractive institutions in the first place. They only pick up the story after institutional formation has already occurred. At best, institutional theories are provisional and incomplete, since they do not explain what's driving the process of institutional formation. They merely look at effects and consequences, making no attempt to ground institutions in a wider causal setting. As a result, institutions in these theories function like theoretical danglers, convenient explanations, or *dei ex machina* that just appear out of nowhere to save the show. Cultural explanations of economic history suffer from the same problems, as we saw above.

To understand why the institutional theory of Acemoglu and Robinson fails to explain anything fundamental about the world, let's consider one of the most prominent examples from the book: Venice in the Middle Ages. Acemoglu and Robinson argue that the initial wave of Venetian prosperity in the Middle Ages was largely driven by inclusive economic and political institutions, such as the *commenda* contracts that provided upward social mobility for young merchants.[63] They posit that the institutional closure of the Great Council in the late thirteenth century, known as *La Serrata*, marked the start of a downward spiral for Venice. From that point forward, only aristocratic elites with hereditary titles could elect the Doge; Acemoglu and Robinson then argue that these elites used their political power to enrich themselves at the expense of other Venetians. And that's basically their story: the rise and fall of Venice was largely about internal and autonomous developments happening within the city-state, as if Venice existed in a vacuum relative to the outside world. Their cloistered historical account manages to do the impossible: tell a story about the rise and fall of Venice without once mentioning the Fourth Crusade, the Byzantine Empire, or the Ottoman Empire. The economic dynamics of Venice in the Middle Ages were almost entirely driven by geopolitics. Writing a historical account about medieval Venice without mentioning the Fourth Crusade is like writing a history of the United States in the twentieth century without mentioning the Second World War. The historian Jason Roche clearly emphasizes the importance of the Fourth Crusade for Venice: "Venice's prosperity in the thirteenth century was largely a product of acquiring Byzantine trading posts in the wake of the conquest of Constantinople during the Fourth Crusade."[64] The Fourth Crusade was the seminal geopolitical event of Venetian history, because the subsequent partitioning of the Byzantine Empire left Venice in control of Crete and other Byzantine territories, thus greatly enhancing its economic and military power. Thousands of Venetians would emigrate to Crete in the thirteenth century, lured by the prospect of huge land holdings, and the island was a major source of lucrative products that boosted Venetian trade.

The rise of imperial Venice is precisely one of the biggest reasons why the Republic's political system became insular: because imperial expansion usually breeds greater concentration of political power in elite hands. One can see this trend everywhere from the Roman Empire to the modern United States. Acemoglu and Robinson buttress their argument of institutional decline with a throwaway statistic about the population of Venice dropping to 100,000 by the year 1500. This number is misleading, first because it ignores that Venice was hit by recurring epidemics that killed many people and forced others to leave the city for different regions within the empire. And second, empire itself is the biggest reason why the city's population decline is irrelevant. By the sixteenth century, Venice was an imperial powerhouse that controlled much of northeastern Italy and dominated Mediterranean trade. The total population of the Venetian empire in 1560 stood at around 2.3 million people, compared to about 3 million for England in the same year.[65] Venice the empire had a far bigger population by 1560 than it had in 1200. Venice the city-state was the linchpin of the entire imperial system, and it would be wrong to think of Venice in the Late Middle Ages as just a city-state. Third, plenty of other European cities experienced massive population declines over the same period. London had 55,000 people in 1520, compared to 100,000 before the Black Death.[66] Was that decline also due to institutional failure? If so, there seems to have been crippling institutional failure across Europe during the Late Middle Ages. That's one explanation; a better one is that it simply took Europe a long time to recover from the demographic calamities wrought by the Black Death and subsequent outbreaks of bubonic plague and other epidemics. The dominant reasons for Venetian decline in the sixteenth and seventeenth centuries can be safely attributed to increasingly ruinous and expensive wars against the Ottoman Empire. Venice lost Crete, Cyprus, and many other important colonial territories in these wars. Although certainly other factors also played a role in the empire's slide toward irrelevance, wars and geopolitical considerations are by far the main culprits.

Geopolitical dynamics are a fundamental driver of institutional dynamics. Political systems can change in major ways as they experience wars, conquests, rebellions, and other violent events.

British history provides many notable examples of this powerful trend. For example, Parliament cemented its dominance over the monarchy largely as a result of the English Civil War, a brutal conflict that left behind a streak of war crimes, a beheaded king, and 200,000 people dead. The creation of the Magna Carta is another powerful example of geopolitical dynamics driving institutional reorganization. In their historical overview of England's economic rise, Acemoglu and Robinson call the Magna Carta a "hesitant step toward pluralism" that they ascribe to the barons standing up to the English king, without ever explaining to the reader exactly where the barons found this incredible courage.[67] The reality is that King John only signed the Magna Carta in 1215 because he had lost a disastrous war against the French in the previous year. That defeat became the latest in a string of English military blunders stretching back over a decade, leading to the French conquest of Normandy. The wars were extremely expensive, and John had raised taxes to help pay for them. Angry that he'd squandered their time and money on a losing effort, English barons revolted and forced John to sign a document in which he ceded most of his control over the English economy. This critical history is ignored by Acemoglu and Robinson, and it's also ignored in most conservative readings of English history, which are so obsessed with emphasizing gradual transformation that they miss the violent and revolutionary moments that set the stage for the later periods of political and economic stability. Another glaring omission in the book is England's loss in the Hundred Years' War, a series of distinct but related wars that lasted more than a hundred years. Defeat against the French permanently turned the English people into a community of islanders. It's a notable example of when losing a war can be a good thing over the long run. Although the defeat eliminated England's continental possessions, it also reoriented the country's strategic priorities toward naval expansion. British leaders gradually relegated the army in importance while heavily investing in the Royal Navy, which eventually became the world's most powerful navy. Another big problem with attributing

England's rise to institutional progress is that much of England's economic advantage was already in place long before the Glorious Revolution of 1688. London's population skyrocketed from 55,000 in 1520 to 475,000 in 1670.[68] England's wool industry also displaced the Flemish and the Italians as the dominant European producer long before 1688.[69]

The fundamental reason why Western Europe became rich first is because of highly favorable ecological dynamics, resulting from the convergence of various geographical, hydrological, geological, agricultural, epidemiological, and other conditions. Acemoglu and Robinson reject ecological explanations of historical development, focusing their fire specifically at the ecologist Jared Diamond, on the basis of a supreme straw man fallacy: that only the geographical distribution of flora and fauna matters for economic development. They argue that many crops and seed varieties critical for agricultural development were found throughout the Eurasian landmass, along with many large mammals that could be domesticated and used for production and transportation.[70] Since all major civilizations in Eurasia had access to these natural resources, they should have followed very similar patterns of development, and yet that's not what happened. That's their argument in a nutshell, but it's based on a highly selective reading of what Jared Diamond wrote in his 1997 book, *Guns, Germs, and Steel*, nor is it representative of what ecologists believe in general. Diamond was clear in the beginning of his book that he wanted to explain variations between Eurasia *and other continents*, a point that Acemoglu and Robinson do acknowledge, not variations within Eurasia itself.[71] And ecologists recognize that economic development can be affected by many other ecological factors besides the geographical distribution of crops and animals. These factors are capable of acting separately or concurrently in space and time, and they include things like viruses, bacteria, volcanoes, earthquakes, droughts, hurricanes, oceanic currents, climatic variations, rainfall patterns, soil quality, and a million other things. Furthermore, complex feedback loops often develop between human societies and the natural world. We're seeing them in action right now with global warming, ocean acidification, water shortages, mass extinctions, air pollution, toxic pollution, and other related phenomena. To think that ecology is not a primary driver of human history is to live in denial. Ecological changes have been strongly implicated in the rise and fall of civilizations throughout history by numerous major books and studies.[72] Examples include the Late Bronze Age collapse, caused primarily by drought and earthquakes, the Classic Maya collapse, caused chiefly though not entirely by drought, and the Germanic migrations into the Roman Empire, caused chiefly by a cooler climate in Northern Europe starting around the year 200.

In his book, Diamond made a useful distinction between fundamental and proximate causes. Institutional explanations of economic development are basically proximate causes; they are provisional explanations contingent on specific environmental constraints. The fundamental causes are the ecological factors lurking underneath. The Industrial Revolution started in England largely because England had highly favorable conditions for initiating energy-intensive industrialization. As I'll explain in greater detail later in this book, these advantageous conditions included factors like vast and easily accessible coal deposits, relatively high wages that gave an impetus to labor-saving technologies, and huge quantities of raw materials obtained from imperialist conflicts and international trade, among other advantages. It doesn't matter how many "institutional changes" the monarchs of Hawaii could make; their society would have never been the first to industrialize under the conditions it experienced. Of course, a new economic or technological process can start somewhere in the world and then be adopted by others, which is totally consistent with ecological theories of economic development. People often bring up Japan as an example that ecology isn't everything, given how rapidly Japan modernized and industrialized despite various ecological scarcities. But this argument is a red herring because Japan modernized by learning from others. The central point is that certain ecological

THEORIES OF ECONOMIC GROWTH AND DEVELOPMENT 145

and material factors are necessary for the *initial* development of a new economic trajectory, but once that trajectory gets going, it's totally normal to expect that it might spread to other places around the world. That's why we have scientific research stations in Antarctica right now. But the Industrial Revolution could never have started in Antarctica! To clarify these and other issues, we need an entirely new theoretical framework for understanding economic development and technological change. Once we have it, we'll be in a much better position to identify what possible paths human civilization should pursue in the future if it wants to ensure economic prosperity, ecological stability, and long-term viability.

CHAPTER 4

The Bionomic Disruption and the Future of Humanity

Life is locked in a seemingly eternal dance with the rest of nature, always trying to stay one step ahead but frequently slipping and falling along the way. Around three billion years ago, tiny organisms called cyanobacteria achieved something incredible and unleashed titanic consequences along the way. They evolved the capacity for oxygenic photosynthesis, learning how to convert sunlight, water, and carbon dioxide into useful energy. It was quite a remarkable discovery because sunlight is such a plentiful energy source; the number of cyanobacteria exploded higher and higher. They were seemingly unstoppable. Oxygen was the main byproduct of all this photosynthesis, going from an irrelevant component of the atmosphere to an increasingly dominant gas. The increased production of oxygen from ocean-dwelling cyanobacteria would end up profoundly changing the future of all life on this planet, because it shifted the distribution of life forms from anaerobic organisms, those that don't use oxygen to produce energy, toward aerobic organisms, which need oxygen to survive.

At first, the vast new supplies of oxygen were likely stored in the oceans, especially ocean floors. About 2.3 billion years ago, however, a series of major geochemical events started transferring large amounts of oxygen from the oceans to the atmosphere, a process that has been called the "Great Oxidation Event" by some scholars and the "Oxygen Catastrophe" by others.[1] The rapid proliferation of oxygen throughout the planet's ecosystems may have set in motion a series of events that culminated in the evolution of complex, multi-cellular life. It also probably led to the planet's first mass extinction event, as many anaerobic organisms could not survive in a newly oxygenated atmosphere. Anaerobic organisms, previously dominant and widespread, would not inherit the Earth. Today, complex life forms that survive without using oxygen do exist, but they are generally minuscule in size. The balance of power gradually shifted toward aerobic life as the ages rolled on. Mammals, birds, lizards, and insects derive from the ancient microbes that managed to harness the power of oxygen in their biochemical reactions.

The "Oxygen Catastrophe" is a searing reminder of the powerful impact that living organisms can have on the planetary ecosphere. It may be tempting to look at the immensity of the Earth, with all its vast oceans and bountiful resources, and conclude that human beings cannot possibly do much to destabilize its ecosystems, and to feel like the planet is largely immune to our civilizational musings. But this attitude is a mirage. We're living in an age when our species is profoundly reshaping almost every aspect of the biosphere. Scholars have called the present age everything from the *Anthropocene* to the *Capitalocene* and *Capitalinian* to signify the unprecedented impact of human civilization.[2] But despite these unprecedented and dangerous developments, not everyone agrees that humanity has anything big to worry about.

Where Humanity Stands

Among many pundits, scholars, and academics, a strong sense of optimism exists about where humanity currently stands. Sometimes the optimism is derived from deep-seated beliefs about the future of humanity. Other times it can arise from the depths of despair, seemingly out of nowhere. When Donald Trump was elected to the presidency in 2016, many liberal reporters and commentators started talking about the climate in irrational and apocalyptic terms that emphasized a disastrous future for humanity. But after Joe Biden defeated Trump four years later, the liberal commentariat shifted its tone dramatically, and full-blown irrational optimism was suddenly in the air, even though hardly anything had changed about the science of climate change as the world kept pumping more greenhouse gases into the atmosphere.[3] The latest waves of optimists have turned into the shock troops of capitalism, arguing that the economic position of human civilization is stronger than ever: global poverty rates are at their lowest levels, global life expectancy has increased substantially, more and more people are receiving a quality education, wars and conflicts are happening less frequently and with lower intensity, and people on average are getting the best healthcare in history. With book titles like *The End of Doom* and *The Better Angels of Our Nature*, the optimists have come to believe that our collective future is an internal affair that can be handled by clever technological tinkering. Their central observation is largely banal but correct: never in human history have so many people, especially in the West and East Asia, been so materially prosperous. But the conclusion that often follows from this observation, that we should keep intensifying capitalism because it has brought us so far, is entirely misguided. It's like telling a sick patient that he should keep taking higher and higher dosages of a risky drug because everything has worked out so far. Although modern civilization has reached incredible milestones over the last two centuries, those achievements guarantee nothing about the future. Quite the opposite is true: the same reasons for the material progress of the modern world in the last few centuries, such as higher agricultural productivity and the burning of fossil fuels, are also the reasons why global civilization is facing the very real prospect of total collapse in this millennium.

Capitalism may have given humanity an unparalleled wonderland of material abundance, but it has done so at a tremendous ecological cost that will inevitably cripple future societies. Our generation is not the only one enamored with wishful thinking and elaborate fantasies. People living during the Pax Romana also glorified the unprecedented period of peace and prosperity that the Mediterranean world enjoyed, but that did not prevent the collapse of the Roman Empire within three centuries of those celebrations. The titanic cost of our achievements is precisely the reason why none of them will last for very long. Humanity is scaling a mountain peak from one side and preparing to fall off the cliff on the other. And the warning signs are all there. Western economies are plagued by stagnation and low growth rates because they are being sabotaged by extractive and plutocratic elites who are capturing all the wealth and income they can. The Chinese hurricane that contributed so much to global economic growth in recent decades is beginning to weaken for the same reasons. Conflicts over precious natural resources have broken out all over Africa and the Middle East, destabilizing national governments and unleashing massive waves of refugees. The optimists of today are not any better than the optimists of previous ages, and what they call "progress" essentially means capitalism on steroids: the systematic exploitation of people and natural resources for the purpose of accumulating more wealth.

Extreme pessimism and outlandish optimism have been reliable staples of the modern age. Both sides have made equally ridiculous predictions about the future. Malthus predicted an imminent population collapse that never came. Some environmentalists in the twentieth century, like Paul Ehrlich, predicted all kinds of imminent demographic and economic calamities for human civilization, none of which materialized.[4] The optimists were hardly any better. Kant imagined that humanity would eventually abolish warfare and reach a state of "perpetual peace" guaranteed by a federation of free states.[5] He certainly did not get his wish with the First World War, which was supposed to be the "war to end all wars." John Maynard Keynes predicted in 1929 that productivity would grow so much in the future that we would all work 15 hours a week.⁶ In an unrestrained orgy of techno-capitalism, the magazine *Popular Mechanics* published a now infamous article in 1950 titled, "Miracles You'll See in the Next 50 Years."⁷ By 2000, according to the article, we would all be flying around in our family helicopters, pollution would be completely eliminated, and no one would ever cook for more than thirty minutes, regardless of how complicated the meals were. For a final gem, the economist Herbert Simon predicted in 1965 that advances in artificial intelligence would make machines capable "of doing any work a man can do" by 1985.⁸ Capitalism has elicited both irrational support and misguided opposition over the years. My critique is not designed to offer specific temporal predictions about the demise of capitalism, but rather intends to show that the intensification of capitalism will present humanity with insurmountable long-term challenges. Nor is the purpose of this work to fan the flames of pessimism. Its goal is to explain, repeat, and emphasize an ancient truth that has been marginalized in modern times: the natural world can have a profound impact on humanity, and humanity can also impact the natural world, for good or ill. Humanity as a species will almost certainly survive the ecological calamities that come in the next millennium, but human civilization stands a real chance of collapsing in the next few centuries.

The idea that modern civilization is even capable of collapse might seem unbelievable to many people. Because we have achieved so much as a species, we instinctively believe that our superiority will somehow carry the day. The same kind of hubris was pervasive among Bronze Age kingdoms and empires over 3,000 years ago, before a series of wars, migrations, and droughts brought most of the states in the Eastern Mediterranean to a swift collapse. The same kind of hubris dominated the thinking of the Maya civilization 1,500 years ago, before a series of environmental changes turned the major Maya cities into little more than ghost towns. Our own contemporary world has witnessed plenty of societies coming to the brink of collapse, including countries like

Yemen, Syria, Libya, and Argentina. The chaotic nature of global capitalism has also managed to severely destabilize previously affluent nations, such as Spain and Greece. Contrary to what the starry-eyed idealists probably imagine, advanced capitalist economies are more than capable of backsliding or collapsing. In the early twentieth century, for example, Argentina had one of the highest standards of living in the world.[9] But the Great Depression and a series of internal political struggles combined to unravel the country's economic fortunes. In the late nineteenth century, Chile had a larger navy than the United States.[10] But just like Argentina, it eventually lost its influential position in the global economy because of external factors that it could not control. Social collapse has been a mundane feature of human history, and it can strike any group of people under the right circumstances. Considering the severity of the ecological crisis we are experiencing, and how much worse things will become, we have more than ample reason to be worried about the potential collapse of global civilization.

Progress is not some rising linear function that stretches back across the ages. Human history is full of ups and downs, of progress and chaos, of discovery and darkness. But the fundamental difference now is that our current "age of progress" has ensnared the whole world, meaning if modern civilization experiences problems that could lead to collapse, these problems would affect every continent on the planet, in contrast to previous waves of collapse that were mostly restricted to specific regions. For the first time ever, we are dealing with the realistic prospect that civilization itself could collapse, not just some societies here and there. The dominant methods of energy extraction under capitalism virtually guarantee that human civilization is heading for the precipice, at least if humanity avoids any radical plan of action to properly manage its economic ambitions. And because capitalism has trapped the entire world in its web, only determined and collective efforts to forge a different path can possibly have any chance of establishing a newer, healthier, and happier economic reality.

But before we get there, it's important to address the concerns and critiques of many dissenting voices on the right and center of the political spectrum. There are many intellectuals who have argued that humanity is doing better than ever when it comes to economic prosperity, that our ecological challenges are real but not significant, and that our current political and economic systems are the best that human beings can achieve. In this chapter, I focus specifically on the writings of Bjørn Lomborg, Ronald Bailey, and Steven Pinker. In his 2001 book, *The Skeptical Environmentalist*, Lomborg argued that environmentalists are raising false alarms over the state of our natural environments and the planetary biosphere. According to him, harmful impacts on the environment have been greatly exaggerated, and one of the main areas where he makes his case is on extinction rates. Contrary to the prevailing opinion among biologists, Lomborg argues that we're not witnessing any mass extinctions right now. Lomborg purports to document "total extinctions since 1600" by looking at data from the Red List of Threatened Species, compiled by the International Union for the Conservation of Nature (IUCN). Based on his analysis of the data, Lomborg claims that "about 25 species have become extinct every decade since 1600."[11] He then estimates an extinction rate that's about an order of magnitude lower than the most conservative extinction rates calculated by leading biologists.[12] Lomborg sums up his position by saying that "actual extinctions remain low."

Lomborg's conclusions are not supported by the evidence. The biggest problem by far is that the Red List is ridiculously biased because it analyzes almost all mammals and birds against conservation criteria but only a tiny fraction of all invertebrates, like snails and insects, are included in its analysis. This omission is inexcusable given that invertebrates constitute the vast majority of all animal life, whether measured by biomass, numbers of species, or raw numerical abundance. Consider the following blatant case. In 2022, a research group estimated that somewhere between 7.5 percent to 13 percent of all mollusks had gone extinct since the year 1500.[13] That would represent something like 150,000 to

200,000 species. By contrast, the Red List gives a figure of 882 on mollusk extinctions for the same period. Another major problem is that the IUCN data is temporally and spatially limited; many species only have data available going back to the nineteenth century and other species are considered only in certain continents or regions. Furthermore, the IUCN has adopted absurdly stringent standards for declaring a species extinct, and it's therefore not surprising that it finds a much smaller number of species going extinct than the actual number that have gone extinct. Finally, there's a powerful reason why nearly all biologists and ecologists agree that the true rate of extinction is much higher than the numbers indicated by the IUCN data, which is that we have simply not documented all the species of animals and plants that exist. In some cases, our knowledge might be shockingly incomplete, judging by the new discoveries of plants and animals that are made around the world every year. This brings up an uncomfortable truth, but a truth nonetheless: we're never going to know anything about many of the species that have gone extinct as a result of humanity's economic expansion over the last few centuries. In sum, it's the overwhelming consensus of modern biologists that humanity is engineering another mass extinction event right now.

Even though the evidence for a sixth mass extinction is rock-solid, extinction is not the only threat facing the biosphere. The ongoing loss of biodiversity is another major crisis for the global commons, given the sheer ecological and economic importance of many wild animals. Insect populations are plunging over much of the world, which is a problem for agricultural productivity since wild insects are highly effective pollinators.[14] For example, the global chocolate industry is worth over $100 billion, and all of that wealth is almost entirely dependent on the pollination of flowers that grow on cocoa trees, a critical function performed by small midges, wasps, and other wild insects.[15] Animals aren't faring much better. A 2018 study estimated that the biomass of wild mammals had declined by an astonishing 83 percent since the rise of human civilization, going from roughly 0.02 gigatons of carbon to roughly 0.003 gigatons of carbon in present times.[16] The same study estimated that wild mammals make up just 4 percent of all mammalian biomass; domesticated mammals constitute about 60 percent and humans represent the other 36 percent, which underscores the sharp collapse in biodiversity since the end of the Ice Age. Interestingly, the study also estimated that chickens currently comprise roughly 70 percent of all avian biomass.

Lomborg's reckless assertions on climate science don't look any better. Lomborg has a habit, both in this book and later in his career, of marginalizing and downplaying the risks of global warming while grossly exaggerating or falsifying its potential benefits. Lomborg claims that the economic costs of constraining global emissions could range "from $3 to $33 [trillion]," which is a huge range filled with uncertainty.[17] He interprets these numbers as the likely range of additional global economic costs after subtracting the cost of doing nothing. And what is that latter cost, you may wonder? When he reports the financial cost that we'd have to bear if we did absolutely nothing about global warming, Lomborg only specifies a single figure of $5 trillion.[18] That's right: Lomborg thinks that if we do nothing about global warming, it'll cost us only $5 trillion and otherwise we're good. Notice how there's apparently no range of uncertainty on the costs associated with doing nothing; it's just a single number! Trying to quantify the specific dollar benefit of saving human civilization is the kind of mental gymnastics that only misguided neoclassical thinkers would attempt. Lomborg's analysis is useless because it doesn't address the costs that global warming will impose on nature, or the various social and political harms and pressures that it would inflict on civilization. An ecological crisis can easily snowball into a political crisis, causing massive waves of war and violence along the way, and potentially leading to a painful social collapse. And as we saw with Nordhaus, estimating the financial costs of global warming on human society is extremely difficult. The reality is that if we do absolutely nothing about our ecological problems and just continue with capitalism as usual, human civilization will inevitably and violently collapse within a few centuries. Of course, our aim is to do something about these problems, but our response needs to be large and ambitious enough to address the corresponding severity of the problems.

A common area of misunderstanding in discussions about global warming is the impact that it will have on mortality during the winter season. In recent times, Lomborg has hyper-focused on winter-related mortality figures, insisting that global warming will reduce winter deaths because these deaths are largely driven by colder temperatures. If the temperatures get warmer, then fewer people will die in the winter. But it's a claim that's based on a false premise, the idea that excess winter mortality is primarily driven by cold temperatures. In reality, most people who die during the winter around the world are *not* dying because they're freezing on the streets or in their basements. The dominant driver of higher winter deaths relative to summer deaths is that many dangerous viruses and infectious diseases spread faster in the winter, for a variety of reasons, including drier conditions that facilitate higher rates of viral transmission and that people spend more time indoors. But cold temperatures are not the only reason why people spend more time indoors, because temperature is not the only major difference between the winter and the summer. Another major difference, of course, is the *length of the day*. In summer days, the Sun is out for much longer than in winter days, which makes it far more likely that people will spend time outside. By contrast, winter sunsets come early in the day and people are therefore more likely to head home early. The seasonal variations in the length of the day are a major confounding factor that complicates any simple causal narratives about temperature changes driving changes in winter mortality. And they're not the only confounding natural or anthropogenic factors that matter; things ranging from humidity and the length of the day to precipitation rates and traffic patterns are all affected by seasonality, so it's extremely premature to attribute all or most excess winter deaths to colder temperatures.

In a 2015 paper, public health specialist Patrick Kinney and his collaborators analyzed the impact of climate change on winter deaths by looking at temperature and mortality variations among thirty-nine cities in France and the United States.[19] The researchers concluded that "most winter-excess mortality is not largely driven by cold temperature." They also held that the evidence is "inconsistent with the notion that future warming would lead to diminished winter season mortality. The lack of correlation between seasonal temperature differences and winter season excess mortality suggests that winter excess mortality is driven mainly by seasonal factors other than temperature." Furthermore, the research group thoroughly addressed and debunked previous studies that projected fewer winter deaths in the future because of increased global warming. Studies that attribute high death counts to cold temperatures generally adopt multi-week moving averages of lagged temperature, and these long moving averages are highly correlated with seasonal mortality patterns, running the risk of multicollinearity and confounding, since it's not certain if the temperature change or the change in season is behind the observed mortality changes. Studies that use smaller moving averages generally attribute lower death tolls to cold temperatures. The researchers also pointed out that "if cold temperatures were directly responsible for winter mortality, one might expect a more pronounced relative winter mortality excess in cities where winter to summer temperature differences are larger, or where winter temperatures are colder." They looked at this issue by comparing Honolulu and Detroit and discovered that the relative seasonal variation in mortality was quite similar even though the seasonal temperature difference between the two cities was roughly an order of magnitude smaller. That's a very unlikely result if cold temperatures are driving excess winter mortality.

Lomborg has made his recent arguments about the effects of global warming on winter mortality by heavily distorting and misrepresenting the results of several studies, chiefly a 2021 study that looked at the correlation between extreme temperatures and global mortality.[20] On a global level, this study's central conclusion was that "from 2000–03 to 2016–19, the cold-related excess death ratio changed by −0.51 percentage points and the heat-related excess death ratio increased by 0.21 percentage points, leading to a net decline of −0.30 percentage points." In an August 2021 article for the *New York Post*, Lomborg inferred from the net decline given in the *Lancet* paper that "climate change saves 166,000 lives each year."[21] But that's a completely ridiculous interpretation. The paper he's citing does not attempt to specifically isolate the effects of climate change on global mortality. In other words, the net decline in cold-related excess deaths could be the result of economic growth, better weather adaptation, improved infrastructure, and dozens of other factors that are all more plausible than global warming. Indeed, one of the authors of that *Lancet* study, Yuming Guo, thoroughly refuted Lomborg's assessment, writing: "It is not correct to interpret that this net decrease was caused by climate change."[22] Guo further notes that he and his team "just estimated the trend of mortality burden related to non-optimal temperatures, but did not do further analysis to examine whether this change is due to climate change or other factors."[23]

Guo further highlights an important point that Lomborg largely ignores: global warming and fossil fuel use have and will continue to have many indirect, complex, and non-linear effects on human health. A great example, as we're about to see in more detail, is that fossil fuel combustion produces dangerous air pollution that kills millions of people every year. Guo explains that "climate change does not only influence temperature-related mortality, but also has other direct and indirect impacts. For example, climate change affects flood, drought, air pollution (including bushfire smoke, sand and dust storms), food supply, and others which are related to increased risks of mortality." He added that "climate change has serious impacts on human health" if we take into account all these factors. Beyond a glib dismissal in his blog, Lomborg had nothing of substance to say about Guo's argument.[24] Indeed, Lomborg kept childishly obsessing over the validity of the net figure, 166,000 fewer deaths, when that was never the core issue. The core issue is whether this net reduction in deaths has been caused by climate change, as Lomborg has repeatedly claimed. The central disagreement is not about the reduction; it's about what caused it. Because Lomborg had no actual evidence that global warming was responsible, he fixated his attention on a red herring.

Lomborg has also questioned whether climate change will make droughts worse, largely relying on poor metrics and incomplete information to make his fragile case.[25] In the topsy-turvy world created by climate change, some regions of the world may experience higher levels of precipitation while others will experience prolonged periods of drought. But overall, there's hardly any doubt that global warming is making droughts more severe. In 2013, the climate scientist Aiguo Dai wrote a landmark paper about the potential for future droughts under global warming. He noted: "The observed global aridity changes up to 2010 are consistent with model predictions, which suggest severe and widespread droughts in the next 30 [to] 90 years over many land areas resulting from either decreased precipitation or increased evaporation."[26] Other papers have also confirmed the results of this work. A 2013 paper from Kevin Trenberth and his collaborators, including Dai, concluded that "increased heating from global warming may not cause droughts, but it is expected that when droughts occur, they are likely to set in quicker and be more intense."[27] In an earlier 2004 paper, Aiguo Dai, Kevin Trenberth, and Taotao Qian concluded that "global very dry areas . . . have more than doubled since the 1970s."[28] Climate experts are in near-universal agreement that global warming has already made and will continue to make droughts much worse.

One of Lomborg's most common tactics to divert attention away from the dangers posed by global warming and fossil fuels is to talk about the millions of people who die every year from air pollution. He engages in this distraction while apparently being unaware that the two phenomena are fundamentally connected. Although the burning of fossil fuels is the dominant cause of recent global warming, it's not the only consequence of modern capitalism's addiction to oil, coal, and gas. Fossil fuel combustion produces lots of harmful byproducts, especially small particles that linger in the air. These particles and many greenhouse gases can be profoundly detrimental to human health, contributing to higher rates of strokes and heart attacks. In 2021, a group of climate and atmospheric science researchers studied the effects of fine particle air pollution from fossil fuel combustion on global mortality patterns. They estimated "a global total of 10.2 million premature deaths annually attributable to the fossil-fuel component of [particulate matter]."[29] After correcting for declining GHG emissions in China from 2012 to 2018, they settled on a new estimate of roughly 8.7 million premature global deaths, happening every year, associated with the burning of fossil fuels. That's an extraordinary figure, and it's much higher than what's often estimated by the World Health Organization.

Other research groups have published similar findings. In a 2020 paper, the atmospheric chemist Jos Lelieveld and his collaborators used the Global Exposure Mortality Model and concluded: "Global excess mortality from all ambient air pollution is estimated at 8.8 million [per] year, with a [loss of life expectancy] of 2.9 years . . . a factor of two higher than earlier estimates, and exceeding that of tobacco smoking."[30] Lelieveld and his colleagues also released a 2019 paper where they looked at the health effects of fossil fuel consumption in greater depth.[31] They showed that curbing fossil fuel consumption could "avoid an excess mortality rate of 3.61 million per year from outdoor air pollution worldwide. This could be up to 5.55 . . . million per year by additionally controlling non-fossil anthropogenic sources." Indeed, fossil fuel use is the dominant reason for the emission of outdoor air pollutants. In its 2016 report *Energy and Air Pollution*, the International Energy Agency stated: "Coal and oil have powered economic growth in many countries, but their unabated combustion in power plants, industrial facilities and vehicles is the main cause of the outdoor pollution linked to around 3 million premature deaths each year."[32]

The report also stated that "coal is responsible for around 60% of global combustion-related sulfur dioxide emissions, a cause of respiratory illnesses and a precursor of acid rain."[33] And furthermore, "fuels used for transport . . . generate more than half the nitrogen oxides emitted globally, which can trigger respiratory problems and the formation of other hazardous particles and pollutants, including ozone."[34]

Red herrings and falsehoods are pervasive among many thinkers on the right, who have made every conceivable effort at defending the status quo. In the book *End of Doom*, Ronald Bailey argues that "democratic free-market capitalist" societies are the only political and economic systems that have solved the problem of long-term sustainability.[35] Citing other scholars, he maintains that pre-capitalist societies were "natural states" that featured "patron-client networks" in which a corrupt elite extracted wealth and resources from an impoverished majority. He argues that elites in such societies were resistant to technological innovation because they thought it would threaten their underlying sources of power. He rattles off a list of such states and civilizations, including the Maya and the Romans, observing that every society on that list eventually collapsed. However, he never really defines what he means by collapse, giving himself maximum flexibility to use the term in any way that he wants. The closest he comes to the task is by suggesting that collapse means a radical shift in the underlying ways of life, often but not always accompanied by large migrations to other regions. This spurious definition is vague enough that it could be recklessly applied to just about anything. For example, he mentions the Soviet Union as a recent example of a "residual" natural state that collapsed. But the breakup of the Soviet Union was not even remotely equivalent to the collapse of the Classic Maya or to the Late Bronze Age collapse. The fall of the Soviet Union did not lead to a mass exodus of people, and some form of central government, however diminished, still survived the ensuing upheaval. Within three decades of the Soviet breakup, Russia had recovered much of its former power, though certainly not all.

Contrast that story with the fate of the great Maya city-states in the Central Yucatan, such as Copan and Tikal, which were completely deserted over a thousand years ago, largely because of conflicts and lengthy periods of drought that gradually crippled agricultural production. Bailey also includes Iraq in the list of modern states that collapsed, completely ignoring that Iraq only "collapsed" because of a brutal foreign invasion in which hundreds of thousands eventually died. Even more damning, however, is what Bailey leaves out of his already shaky collapse narrative. Nowhere does he mention the collapse of the British and French colonial empires, both of which were critical to the emergence of capitalism. In the early twentieth century, the British Empire controlled roughly 25 percent of the world's land.[36] By the end of the century, Britain was left with its core territories in Europe and some small, scattered islands around the world. Most sane people would say that these events represent a collapse of some kind or another, but Bailey does not even bother mentioning it. He can't acknowledge such inconvenient facts because they would destabilize his carefully orchestrated propaganda.

Let's revisit Bailey's exact definition of a so-called natural state: "hierarchical patron-client networks in which small, militarily potent elites extract resources from a subject population."[37] Consider that this definition could be seen as an almost perfect description of the modern United States, the world's leading superpower. The United States has a larger military budget than the next five highest spenders, operates hundreds of military bases and installations around the world, extracts money from host nations to finance its military presence, places severe economic sanctions on those who challenge its authority, and leverages the adoption of the dollar as the world's de facto reserve currency in order to borrow cheaply from other countries. Amplifying these mistakes, Bailey also makes the glib assumptions that democracy and capitalism are actually compatible and that capitalism has somehow avoided extractive development. Both of these claims are bogus. Research studies have suggested that modern economies are the most unequal in history, at least when we look at wealth distribution.[38] The emergence of Western capitalism relied heavily on forced and uncompensated labor from tens of millions of people around the world.[39] And plenty of reputable economists and political scientists have rightfully concluded that Western nations resemble plutocracies more than democracies.[40] In other words, the rich rule in the contemporary West, not the people. The political organs of the modern nation-state have become decorations on a Christmas tree: pretty to look at, but easy to switch on and off as the ruling classes demand. Bailey conveniently forgets that private capital has long indulged in a torrid love affair with a distinctly anti-democratic system: fascism. Capitalists absolutely love a strong, powerful, and centralized state that crushes labor unions and egalitarian political movements. It feeds into their ultimate dream: a servile workforce that provides cheap labor without making a fuss. In his 1927 work *Liberalism*, the Austrian thinker Ludwig von Mises, whose ideas in many ways inspired the modern libertarian movement, flatly stated: "it cannot be denied that fascism and similar movements aiming at the establishment of dictatorships are full of the best intentions and that their intervention has, for the moment, saved European civilization. The merit that fascism has thereby won for itself will live on eternally in history."[41] Less than twenty years after this precious gem was published, fascism lay in ruins all over Europe at the end of the Second World War. The dalliances between capitalism and fascist dictatorships are almost too numerous to recount, and they persist to the present day. From Hitler and Mussolini to Franco and Pinochet, capitalism and fascism were a match made in heaven.

Capitalism has not been a friend to democracy, nor has it produced a healthier environment, as Bailey claims. As we've already seen, richer nations look cleaner and greener to a large extent because they have transitioned energy-intensive manufacturing to other poorer regions around the world, and also because they've simply fabricated and invented whatever numbers are necessary to show environmental progress. Many rich nations eliminate a substantial portion of their trash and recycling waste by selling it to poor countries, often in violation of international standards and agreements.[42] Capitalism has simply shifted the geographic locus of ecological degradation, and in so doing it has laid the groundwork for upcoming disasters, such as intensified global warming and mass extinctions. Capitalism has only reinforced and magnified the sins that Bailey claims to oppose. Bailey's explanatory framework relies on the debunked idea that the right cultural values and sociopolitical institutions have turned Western capitalism into the most productive economic system ever. By now, however, historians have documented multiple cases of civilizations that allowed for greater market freedoms than European nations before the nineteenth century. Under the Song, Ming, and Qing dynasties, China had vibrant and largely decentralized markets where merchants from all over the country could gather and exchange goods. During the Qing years in particular, a merchant could deposit hard currency into a bank, receive a remittance certificate specifying the deposited amount, and would then be able to exchange this certificate for goods and services in virtually any market.[43] This kind of freedom in mobility and exchange was generally forbidden in most European states. The Chinese government did intervene in certain critical sectors, like the salt industry, but the level of intervention never approached that of its European counterparts, which were heavily involved in the development of capitalism, to the point of launching numerous wars on behalf of their corrupt aristocracies.

If anything, the rise of capitalism was accompanied by interventionist states that understood the importance of political power and military force in shaping global markets and trading systems. As the notorious Dutch imperialist Jan Coen once said, "There can be no trade without war, and no war without trade."[44] Bailey does not recognize the critical importance of warfare in the development of capitalism. He hardly says anything about it, which is not surprising. He imagines markets as these divine and metaphysical realms where all sorts of magical things happen, rather than seeing them for what they are: methods and structures of economic transactions that are shaped by various people interacting together, sometimes peacefully and sometimes violently. The fundamental reasons for the incredible leaps in technological development experienced by the West have virtually nothing to do with values and institutions. This is exactly the kind of idealist, nostalgic propaganda that makes historiography lazy and boring. Values and institutions can influence certain features of economic systems, but material forces are the causal pumps that irrigate the fields of civilization. Our thoughts and ideas are material systems dynamically coupled to external economic conditions, including the technological products that shape our lives. Capitalism took off in the Western world because of favorable material and ecological conditions that converged at the right time. European values and institutions were secondary effects of this larger historical process.

Not all defenses of the status quo are about ecological sustainability. Some thinkers, like the psychologist Steven Pinker, are enamored with the current order because they believe that it's made humanity far more peaceful. In his 2011 work, *The Better Angels of Our Nature*, Pinker claimed that humanity in modern times has become more peaceful compared to our ancient and medieval ancestors. To back up his claims, he rattled off a series of statistics, tables, and charts that supposedly indicate declining levels of violence in modern history. Pinker is part of a generation of liberal scholars who have convinced themselves that things are better now than they've ever been before. The "democratic peace theorists" in this school of thought believe that much of humanity's peaceful streak after the Second World War can be explained by the spread of liberal democracy around the world; they credit liberal democracy for encouraging more substantive political deliberation and therefore making it less likely for democracies to go to war against each other.[45] It's a seductive idea; after all, who doesn't want the better angels of our nature to overcome our worst demons? But for all his pretensions to using data to back up his claims, much of Pinker's evidence about the history of warfare is useless in fundamental ways.

Indeed, it's very easy to show that Pinker's central claims are false and misleading. In his book, Pinker presents a table that contains a list of the most violent conflicts or events in history, which includes conflicts like the Second World War.[46] He then shows another table of the deadliest conflicts in which he has modified the numbers to account for things like world population and other statistical factors.[47] After these changes, he concludes that the An Lushan Rebellion from the eighth century was the deadliest conflict in history, while placing the Second World War just *ninth* on the list. These two prominent cases highlight some major problems with his overall argument. First, death tolls from ancient wars are frequently unreliable. Ancient writers and chroniclers were notorious for inventing ridiculous casualty figures, leaving modern historians to pull the pieces together and come up with more realistic numbers, all of which are sketchy and tentative. Even many modern wars have conflicting death tolls, depending on the available evidence, ideological preconceptions, and even on what people count as war-related deaths. There's no broad agreement, for example, on the number of troops that Ukraine and Russia have lost in their recent war, and that's a conflict unfolding in the age of social media. The data quality on most historical conflicts is so poor and unreliable that it's practically useless to perform per capita death toll comparisons between ancient and modern conflicts.

But the more fundamental problem is that Pinker blindly cites data points that he doesn't understand in the slightest. The An Lushan Rebellion in eighth-century China under the Tang dynasty is sometimes said to have produced about 36 million deaths, which is the number that Pinker specifically mentions in his unadjusted table. This death toll is estimated on the basis of a post-Rebellion census administered by the fragile and recovering Tang government. But modern historians of the period understand perfectly well what happened: the weakened and bankrupt Tang imperial court simply did not have the resources to send out a large number of officials, hence the census resulted in a severe undercounting of the actual population. In her 2000 book, *The Open Empire*, the historian Valerie Hansen writes:

> During the years it took the court to put the rebels down, the government had neither the manpower nor the funds to carry out the triannual land and population surveys the equal-field system demanded. The number of registered households in the Tang Empire dropped from nine million in 755 to two million in 760—not because the population diminished but because the system of household registration was not enforced.[48]

This was not just a problem with this rebellion either; it happened repeatedly throughout Chinese history. For a similar example in the same century, according to Richard Guisso in the *Cambridge History of China*, only three million households were registered under the reign of Emperor Taizong from 626 to 649. Under the Sui dynasty, by contrast, nine million households had been registered. Guisso says that such a "sensational decline was not the result of catastrophic loss of life during the civil warfare by late Sui and early Tang, but of simple failure by the local authorities to register the population in full. . . . Considerably more than half the population was thus unregistered and paying no taxes."[49]

The point is that the number of people who actually died in the Rebellion was *far fewer* than 36 million. It's interesting to note that Pinker cited Michael White, a librarian who studies casualty figures from warfare, as a source for the figure of 36 million deaths. But in another 2011 book, which actually had a foreword from Pinker, White says that he believes a better estimate of the death toll is 26 million, and finally settles on the number 13 million to be more "conservative."[50] Pinker couldn't even properly cite details from his own sources. We're never going to know the actual number of people who died in the An Lushan Rebellion, but considering the localization and intermittency of the war, it's a safe bet that the real death toll was *far less* than 5 million, implying that the An Lushan Rebellion is *not* actually the deadliest conflict in history, even by Pinker's statistical adjustments.

Now let's consider the Second World War. There are often two prominent sets of death figures cited for the deadliest war in history. Values in the first set, the more popular one, range anywhere from 50 to 60 million total deaths for the entire war. Pinker falls in this camp, citing a figure of 55 million deaths. Values in the second set usually range from 70 million to 80 million deaths.[51] That's quite a big gap for a war that happened relatively recently in historical terms. What explains it? Basically, it comes down to what people want to consider as a "war-related death." The first set only counts deaths that resulted from fighting and violent interactions, up to and including aerial bombing campaigns on civilian populations. But the second set also includes deaths from diseases and famines that can be indisputably traced back to the war itself. The Bengal Famine in British-occupied India would be a prominent example of a nonviolent event that was fundamentally caused by violent conflict and by Britain's decision to prioritize resources for its forces in the European and African theaters. Given these two different perspectives, what should we take as the total death toll of the Second World War? Was it about 55 million or about 75 million? Reasonable people can disagree. The more important point is that this analysis raises some philosophical problems about causality. Wars have many complex and chaotic effects beyond the immediacy of the violent confrontations. To what extent should we try to measure these more chaotic, lingering impacts? One could plausibly argue that they're worth including in final death toll figures if we want a more complete assessment of all the harmful effects caused by warfare. If so, this position would virtually negate Pinker's premise that modern wars have produced far smaller per capita deaths, because it would have the effect of *dramatically* increasing the death tolls from most modern wars, including both the First and the Second World War.

As if all these flaws weren't bad enough, there's yet another critical problem with Pinker's thesis: the problem of demarcation, meaning the problem of what counts as a war in the first place. Pinker has made some highly dubious and misleading choices when he quantifies the number of wars. For a galling example, in the table where he ranks the deadliest conflicts in history, he categorizes all the Mongol wars of the thirteenth century as *a single war*; he does the same thing for the wars associated with the collapse of the Roman Empire, a period spanning over two centuries. Here's why this sleight of hand matters. Suppose that 40 million people died in this "single" Mongol war and that the global population was 400 million. Then the per capita death toll from this war would be 40 million divided by 400 million, or 10 percent. Next, imagine a scenario where 40 million people also died, but over *five wars* instead of one. Now the average per capita death toll from all the wars, assuming a uniform distribution of deaths, would be 8 million divided by 400 million, or 2 percent. In other words, per capita death tolls from war are *highly sensitive to how wars are counted and aggregated*. Pinker lumped very complex wars in the past into a single basket, which has the effect of artificially boosting per capita death tolls from older wars, thereby making per capita death tolls from modern wars seem lower by comparison. He's not the only one who engages in these statistical gimmicks. For example, the researcher Max Roser created a chart called "Global Deaths in Conflicts Since the Year 1400," which circulated widely among various news sites. His main source for premodern war deaths was the Conflict Catalog, which also lumps many major wars into a single category. For example, the Conflict Catalog treats the Hussite Wars in the Middle Ages as one war, even though in reality they involved multiple distinct crusades and other complicated conflicts, and the per capita death toll averaged across all these different wars would be much lower than the per capita death toll from all the wars combined. Likewise, the Conflict Catalog lumps all of the complex wars, battles, and treaties of the Hundred Years' War in "Phase 3," lasting from 1415 to 1444, into a single bucket.[52] It's curious to note that, even with these flawed aggregation methods, Roser's chart showed the middle of the twentieth century as having the highest death rates, but just barely. However, had older wars been shown as the distinct conflicts they actually were, then death rates in the twentieth century would be vastly higher than anything else seen over the past six centuries.

In short, there is no reliable evidence that humanity has become more peaceful in modern times. But if democratic peace theory and the "everything is awesome" crowd are wrong, then we still have the problem of explaining the cause for the relative drop in violence following the Second World War. There is indeed something that requires an explanation here. Why don't Italy and France go to war anymore like they once did? Is it because they're "democracies" that have learned to respect each other? Not quite. The reason they don't go to war is because they're puppet regimes of the United States, and they lack sovereignty and independence in their foreign policy. Since the United States is the prime hegemonic power in Western Europe, it effectively dictates what most European nations can and can't do when it comes to their external relations with other countries. Italy and France avoid war with each other for the same fundamental reason that Athens and Sparta didn't fight each other for centuries under Roman rule: they are satellite regimes of a bigger power, and they don't control their strategic destiny. On almost all big international issues, they have to do what the United States wants. Call this the *imperial dynamics theory*, which goes something like this. States expand and become powerful either through conquest or through establishing spheres of influence; these two activities can overlap. They then consolidate their gains by suppressing internal dissent and competition among dominated regions; think of the stability enjoyed by the Mediterranean world under the Roman Empire, the Pax Romana, or the current peace that Western Europe is enjoying under American domination, the Pax Americana. The consolidation phase is characterized by challenges and conflicts along the imperial periphery, which lies just beyond the sphere of domination, hence Rome incessantly at war with Persia and the United States incessantly at war in the Middle East. Internal problems and strategic overextension combine to unravel the imperial system, thus destabilizing the international balance of power and encouraging new conflicts to fill in the resulting power vacuum, which is what happened with the wars that accompanied the fall of the Roman Empire or the wars that followed the fall of the Soviet Union between Armenia and Azerbaijan, Russia and Ukraine, and other combatants. For an obvious example, Greece and Turkey would have been at war a hundred times over in the past three decades if it weren't for the fact that they're both American client regimes. It's the big sharks that keep their little sharks in check, while at the same time preying on other little sharks; once the big sharks go away, the little sharks start to squabble until one of them becomes the new big shark.

Humanity and the Biosphere

Our destiny is tied to the natural world. It has quite the impact on us and we've certainly had quite an impact on it. We're living through an age of profound ecological disruption, the likes of which are hard to discern in the planet's recent history. But it's not the only time that life on Earth has been tested. Despite multiple mass extinctions, crashing asteroids, continents uniting and breaking up, volcanic eruptions, ice ages, and so many other adverse conditions, life continues to hang on and thrive. After billions of years of change and evolution, the total biomass in the biosphere has reached about 550 gigatons of carbon.[53] Plants, the dominant kingdom with around 450 gigatons, are largely terrestrial. Animals, comprising roughly 2 gigatons, live mostly in aquatic environments. Bacteria, about 70 gigatons, and archaea, roughly 7 gigatons, mostly inhabit deep subsurface environments.

Modern civilization has been ravenous in its expansion and energy use, but human beings have been changing the biosphere for thousands of years, long before the emergence of airplanes and the internet. For example, human activity contributed to the Quaternary

Megafauna Extinction between 50,000 and 3,000 years ago, which claimed around half of the large land mammal species, including the woolly mammoth and the saber-toothed tiger. In 2011, the energy scientist Vaclav Smil analyzed the natural resources extracted by human beings during the past 5,000 years.[54] Specifically, he analyzed the amount of phytomass consumed by human beings, along with other extractive activities that impact the biosphere. Phytomass is the amount of mass contained in plants, organisms that use photosynthesis for their energy needs. Because plants comprise the vast majority of all living things by weight, phytomass is an important proxy for how the planet's overall biomass is doing. Smil noted that the shift from subsistence foraging to settled communities started happening after the last glaciation event, which was followed by a huge expansion in phytomass.

The earliest civilizations had limited capacities for energy use, often resorting to burning wood and crop residues. But more complex societies eventually mastered agriculture by planting crops and domesticating large mammals for tilling and transportation. These developments drove higher per capita energy use and the greater exploitation of natural resources. Human activities over the past 2,000 years, according to Smil, may have reduced the stock of phytomass by up to 45 percent. Over the twentieth century alone, global phytomass declined by roughly 17 percent compared to the total in 1900. Global anthropomass, the total mass of all humans, more than quadrupled over the twentieth century, reaching roughly 55 megatons of carbon. By 2000, the global anthropomass was about ten times larger than the global zoomass of all wild terrestrial mammals, which stood at about 5 megatons of carbon, according to rather liberal estimates. At the beginning of the twenty-first century, the global harvest of food, feed, and fiber crops amounted to roughly 2.7 gigatons. Residues and forage crops pushed the total of aboveground phytomass available for harvest to about 7.6 gigatons. Roughly half of this phytomass was fed to animals, and it produced about 300 megatons of meat and about 700 megatons of milk.

Human activity has also profoundly impacted forests, the dominant terrestrial ecosystem on Earth. Forests cover about 4 billion hectares globally, or roughly 30 percent of the planet's total land area.[55] They also hold about 92 percent of all the planet's biomass, so the distribution of forests is tantamount to the distribution of biomass. Roughly 5 percent of forests globally are actually plantations used for commercial purposes. After the last glacial maximum 18,000 years ago, global carbon storage in vegetation and soil roughly doubled. Biomass in natural vegetation peaked about 10,000 years ago, which is around the time that agriculture started taking off in many parts of the world. And once it did, things were never the same; human consumption and exploitation of biomass products has cut the total global biomass roughly *in half* relative to the recent peak level. Global environmental changes caused by human economies can exert complex effects on forest productivity and carbon storage. It's been estimated that approximately 13 percent of global anthropogenic carbon emissions from 2000 to 2010 resulted from tropical net deforestation.

These trends have changed slightly in recent times, but not in any significant way that would change our current ecological trajectories. Since 1990, biomass density has increased in established forests worldwide, suggesting that extra carbon dioxide in the atmosphere may be enhancing biomass gains. Satellite data from 1982 to 2016 has revealed that global tree cover has increased by 7 percent relative to 1982 levels.[56] The net gain was largely driven by huge expansions of temperate and boreal forests, especially in the Northern Hemisphere. However, the tropics experienced severe deforestation over the same period, collectively losing about 900,000 square kilometers of tree cover since 1982. That's important because tropical forests account for two-thirds of all terrestrial biomass and are a major carbon sink. A 2018 study from Global Forest Watch found that tropical deforestation accounts for 8 percent of the world's annual carbon dioxide emissions.[57] That makes tropical deforestation a bigger contributor to global warming than the entire European Union.

THE BIONOMIC DISRUPTION

Human civilization now stands at a critical juncture. In 2009, a team of ecological scientists published a major framework focused on defining critical planetary boundaries that human civilization should not transgress if we want to continue functioning safely. They wrote:

> Although Earth has undergone many periods of significant environmental change, the planet's environment has been unusually stable for the past 10,000 years. This period of stability, known to geologists as the Holocene, has seen human civilizations arise, develop and thrive. Such stability may now be under threat. Since the Industrial Revolution, a new era has arisen, the Anthropocene, in which human actions have become the main driver of global environmental change. This could see human activities push the Earth system outside the stable environmental state of the Holocene, with consequences that are detrimental or even catastrophic for large parts of the world.[58]

Their proposed boundaries cover a variety of systems and processes that are essential to the wider biosphere, including climate change, the rate of biodiversity loss, the nitrogen cycle, the phosphorous cycle, ocean acidification, chemical pollution, and global freshwater use, among others. For example, they recommend that human beings try to limit the atmospheric concentration of carbon dioxide to 350 parts per million; back in 2009, that number was roughly 387 parts per million, and as of 2023 it had shot all the way up to roughly 420 parts per million. Crossing a threshold for a short period of time does not mean doom for humanity, far from it. We're already past their carbon dioxide boundary and human civilization hasn't collapsed yet. But transgressing many of these planetary boundaries for long periods of time would most definitely threaten the viability of global society, especially because it would substantially increase the risk of triggering ecological tipping points and critical biospheric thresholds, like massive disruptions in ocean circulations and rapid sea level rise. The specific thresholds they chose can also be questioned and debated, but the overall framework is useful because it offers human societies concrete action plans for managing the global ecological commons.

And the global ecosphere is now in a state of peril. The early phases of the ecological crisis have already arrived. In 2017, Puerto Rico was struck and heavily damaged by a powerful hurricane lurking over unusually warm waters, sending much of the island back several decades economically. That same year, a historic drought in Argentina crippled agricultural exports and triggered a massive recession, which eventually coupled with a currency crisis and forced the country to borrow billions from the IMF for the second time in less than two decades.[59] Severe and unusual droughts in Central America are also disrupting agricultural production and playing a major role in convincing hundreds of thousands of migrants to head north.[60] Major droughts and water shortages in Afghanistan have fueled widespread resentment against the central government in Kabul and have incited tensions between the country and its neighbors.[61]

In the summer of 2022, a lack of rain and record temperatures in western and central China created historic drought conditions that brought much of the country's economy to a screeching halt.[62] The Chinese economy is heavily reliant on hydroelectric dams, so when the rains don't come, electricity production in much of China collapses, forcing widespread closures in commercial and industrial districts. Many factories in Sichuan province had to be shut down for a week or more. Some river levels dropped so low that ships could no longer move. On the other side of Eurasia, heat waves in France caused several rivers to become warmer than usual, making it more difficult to cool down some of the country's nuclear reactors and forcing their temporary shutdowns.[63] In Pakistan, roughly one-third of the country's total area was flooded by massive rains in 2022, causing major disruptions to the country's economy.[64] These and thousands of other simultaneous developments are only the opening lines in a multi-act play that human civilization will nervously witness and experience over the next few decades and centuries.

The world's leaders have started to notice that we're entering a period of ecological upheaval, but concrete actions and meaningful responses have been lacking so far. In 2016, the International Resource Panel concluded that the global consumption of materials since 2000 had grown at a faster rate than global GDP, adding that "global material efficiency, for the first time in a century, has started to decline."[65] In 2017, the chief economist of Norway's oil giant Equinor, Eric Waerness, wrote that decoupling economic growth from energy consumption "might be impossible."[66] In 2018, a major report from the International Panel on Climate Change stated that preventing catastrophic levels of global warming would require "rapid, far-reaching, and unprecedented changes in all aspects of society."[67] Antonio Guterres, the Secretary General of the United Nations, told a climate conference in early December 2018 that "we are in deep trouble with climate change."[68] Optimism has finally given way to realism, even if many of these individuals and organizations fail to notice the next required step: a full-blown social, political, and economic transformation of our societies. The next eight chapters of this book are devoted to precisely this transformation: explaining the theoretical and empirical justifications for it and then outlining an ambitious action plan for the future of human civilization.

CHAPTER 5

Energetic Conversions in the Economic Process

Energy is embedded in all human actions. But it's not something that's just lying around, ready to be used. All conceivable economic transactions, from the exchange of money to the production of commodities, require energetic *conversions* from various sources. Oil is first extracted from the ground, then converted to gasoline at a refinery, and finally the gasoline is burned by your car's engine, which converts the chemical energy of the gasoline into the mechanical work of the tires. And while you're busy browsing the internet, the solar panels are taking the light energy from the Sun and converting it into electricity, which is then sent to your home so you can stay online. The conversion of energy into different forms is what makes civilization possible. It's what allows us to do things like drive to the grocery store, surf the web, play video games, watch television shows, and read romance novels at the beach. In this fundamental sense, all economic activities depend on energy flows, and none of those activities can exist separately from the laws of physics. Our economic possibilities are constrained by physical and environmental conditions. We're going to begin our theoretical journey by understanding the nature of these constraints, specifically by studying how the conversion of energy impacts the organization of society, and also how the organization of society affects the conversion of energy.

Although it's tempting to see economics as just physics in motion, we should strongly resist this urge toward naïve reductionism. Economics is about a lot more than physics and energy. We live in a *social* world of money, symbols, languages, stories, ideologies, classes, hierarchies, and institutions. All of these critical features of human existence affect the structural organization of our energy networks. What's really important to understand about our economies is not just biophysical facts, like how much energy they're using or how much they're dumping to the environment. It's also important to know how the complex interactions among different classes and social hierarchies collectively organize the productive activities that require those energy flows. These social aspects of our existence strongly constrain the dynamical evolution of energy flows, and that implies something important: if we want to construct a society that can be compatible with its natural environment in the long run, an effective way to do it is precisely by changing the classes and hierarchies that are responsible for the incompatibility in the first place. Nevertheless, for now I will largely though not entirely keep the focus on the biophysical basis of economic activity, before comprehensively addressing the emergent social issues later in this book.

Exergy and Efficiency

Heat engines, like car engines or steam engines, acquire energy from a hot reservoir, use a portion of that energy to do useful work, and dissipate the rest of the energy to a cooler environmental sink. Because of dissipative effects like friction, real-life engines cannot entirely convert all of the energy they consume into useful work, which gives rise to a notion of *energy efficiency*: the useful energy output divided by the total energy input for a thermodynamic process. The most efficient heat engine possible is known as a *Carnot engine*, after the French engineer Sadi Carnot. In the nineteenth century, Carnot discovered that the maximum efficiency any heat engine could have equals 1 minus the ratio of the cooler temperature of the environmental sink to the hotter temperature of the energy reservoir.[1] If an energy reservoir at a temperature of 500° Kelvin is delivering energy to an engine that's dumping the wasted energy to an environment with a temperature of 300 Kelvin, then the maximum energy efficiency of the engine is 40 percent. In other words, if the engine worked "perfectly," it could convert 40 percent of the total energy consumed into useful work.

This upper efficiency limit implies that, in any thermodynamic process, there's a constrained amount of energy available for useful work. This maximum amount of available energy that can be obtained from some physical process is known as *exergy*.[2] Unlike energy, which is always conserved in total, exergy is something that changes over time. The exergy of a physical system declines as its entropy rises, which is just another way of saying that increasingly disordered systems generate less and less useful energy. Figure 5.1 shows the available exergy in the United States, and the *actual* useful work produced with that exergy, from 1900 to 2000. A rough proxy for the collective *exergy efficiency* of the United States can be calculated by taking useful work and dividing it by exergy. In short, the exergy efficiency is defined as the useful energy output divided by the maximum energy available for producing useful energy. As the plot amply demonstrates, the vast majority of the available exergy in the United States was never converted into useful forms of energy. In 1900, the aggregate efficiency stood at about 4 percent. By 2000, it had increased to 11 percent, a 7-percentage point gain. But that figure still implies that almost 90 percent of America's available exergy in 2000 was lost as waste to the environment.

To have a slightly more quantitative understanding of the differences between these two types of efficiencies, let's consider two steam engines that each consume 1,000 joules of energy in every

ENERGETIC CONVERSIONS IN THE ECONOMIC PROCESS 179

FIGURE 5.1: Total Exergy and Useful Work in the United States, 1900–2000, Measured in Terajoules

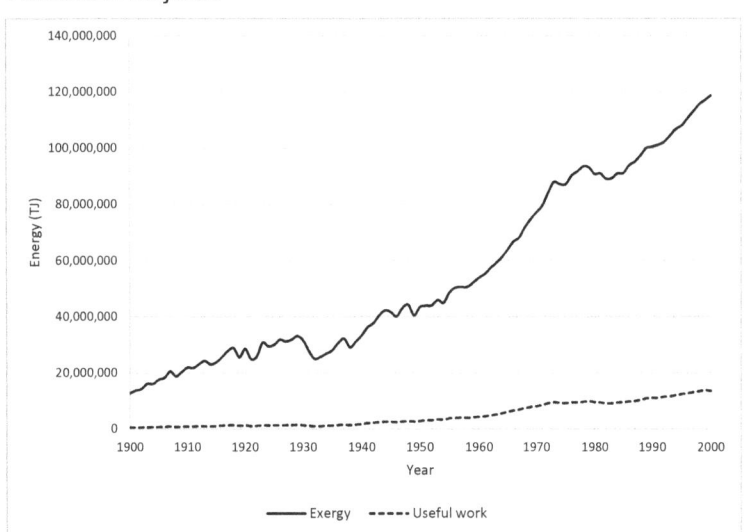

Note: The unit of energy for this plot is the *terajoule*. A terajoule (TJ) is equal to a trillion joules, or 10^{12} joules.
Source: The data is obtained from Benjamin Warr's Resource Exergy and Useful Work Database, https://sites.google.com/site/benjaminwarr/the-economic-growth-engine/rexs-database.

cycle while producing 400 joules of useful mechanical work. The energy efficiency of both is 40 percent, or 400 joules divided by 1,000 joules. Suppose that the first engine is operating at a temperature of 600° Kelvin and dumping its waste losses at 300° Kelvin, which is roughly room temperature. The maximum efficiency of this engine is 50 percent, so only 500 joules of the total 1,000 joules are available to do useful work. Since we got an actual output of 400 joules, the exergy efficiency of this engine is 80 percent, or 400 divided by 500. Now suppose that the second engine is operating at a temperature of 1,000° Kelvin and dumps its heat losses into an environment of 300° Kelvin. The maximum efficiency of this engine is 70 percent, so 700 joules of the original 1,000 joules are available for useful work. Because this engine also produced 400 joules of useful energy, its exergy efficiency is about 57 percent, or 400 divided by 700. Even though both engines have *the same energy efficiency*, their *exergy efficiencies* differ substantially because the first engine is much better at converting its available energy to useful work.

Both of these definitions have their advantages and disadvantages. Exergy efficiency is useful for understanding how to improve our industrial and technological capacities. It's a way of measuring how efficiently we're using the energy that's *actually available* for useful tasks. On the other hand, energy efficiency is more useful if we want to understand the ecological impact of our economic activities, because it actually measures the total losses dumped into the environment from the physical processes powering our civilizations. At an aggregate level, they both highlight important though different features of our energy-intensive civilizations. Efficiency metrics like these are examples of thermodynamic efficiency, in the sense that they're measuring thermodynamic quantities like heat and work. But that's not the only kind of efficiency people care about. A company shipping goods across the United States might be more interested in the metric of cost per unit weight, because it's better for the bottom line if it can ship more stuff at cheaper prices. Based on that metric, the company might decide to ship goods using trains on railways instead of trucks on highways, since unit shipping costs are lower for rail than other forms of ground transportation. The company probably doesn't care about the thermodynamic efficiency of the truck engine versus the train engine, unless the differences are so astounding that they have a major effect on shipping costs.

Indeed, the ecologist Murray Patterson pointed out various problems with how efficiency is traditionally defined.[3] For example, the standard thermodynamic definition of efficiency, in which useful energy output is divided by total energy input, implicitly assumes some kind of value judgment because we need to decide what's useful in the first place. Another issue is the quality problem. Patterson points out that traditional indicators of energy efficiency don't account for the differences in low-quality energy sources like coal and high-quality sources like electricity. Because electricity is a highly refined and concentrated form of energy, it can readily and quickly provide end-use services, such as powering your computer or television. But raw coal hasn't been converted into anything useful yet; it cannot deliver end-use services in its current state, hence it has lower quality compared to electricity. A useful analogy here is food. A bunch of apples and a bag of potato chips may have the same number of calories, but the nutritional value of the apples is far better. Because standard efficiency indicators don't account for quality differences, they may give a misleading picture of an energy system's actual performance. Methods have been developed to account for these quality differences, but it should be emphasized that the quality problem is itself dependent on the valuation problem, as human judgment is what's deciding whether something is a low-quality or a high-quality energy source in the first place.[4] The basic point is that the term *efficiency* can acquire various physical, industrial, economic, and cultural connotations depending on the context in which it's applied and who's using it. But here our focus is on efficiency in the sense of which initial stock of energy can be converted to a different form that can do something useful, according to the prevailing social conventions of what's useful.

The Conversional Network

The conversion of energy is a complex process that drives every aspect of civilization. A naïve intuition is that energy makes its way around the economy in a kind of linear chain. The resources are first extracted from nature, then they're changed into raw materials, then the raw materials are shipped to the factory, then the people and the machines in the factory convert the raw materials into a finished product, and finally the manufactured product is delivered somewhere and sold to consumers, who use it for various purposes in their daily lives. We're going to rely on a better metaphor for how societies actually use energy: a spatial and temporal *network* in which different nodes are exchanging energetic signals with one another. The nodes can be anything that converts energy: steam turbines, airplanes, photovoltaics, human bodies, residential homes, commercial businesses, and even entire economies or countries at larger scales of analysis. The concept of a node is designed to be flexible enough that all of the systems above can be seen as networks themselves, with their own constituent nodes and signals. When energy moves around the economy, seeing it as an organized network emphasizes the *holistic interdependence* between the economic system's different parts. Conversions are happening at every single node, as well as in all the logistical channels that link the nodes. Conversional networks—I often call them *coronets* for short—are biophysical pathways through which energy is changed or converted into different forms. The brain is an example of a conversional network: it takes inputs like light waves and sound waves and converts them to electrochemical signals activated across billions of cells. Power plants and electrical grids are another example: they take inputs like fossil fuels and renewables and convert them to electricity, which is then transported over long distances and further transformed in various ways before supplying homes and businesses. Indeed, an entire economy can be thought of as a conversional network, since it also takes various inputs and converts them to useful outputs like electricity and mechanical work.

A conversional network includes several stages of extraction, production, distribution, and consumption, all collectively interacting via logistical channels, which can be things like power lines, highways, pipelines, and railroads. These stages and methods require a stream of interdependent conversions. Think about the extraction of oil. To pull oil out of the ground, we first need to get drilling equipment, along with other vehicles and machines, to the location where the reserves are found. If the machines malfunction, they'll need to be repaired or replaced, and that whole process requires energy. After extracting the oil, companies then send it to a refinery so it can be converted into something useful— fuel for cars, let's say. The extraction point is the spatial node into which multiple logistical channels are fed, and the same can be said for the production and consumption nodes as well. Conversional nodes are tightly coupled via their logistical channels, which means that a disruption anywhere in the network could resonate and reverberate throughout the entire system. But not all parts of the network are equal. Although every single part of the system runs on energy conversions, some parts are more important than others because they act as gateways or chokepoints for the additional activation of other parts. Every coronet has critical energy hubs, like power plants and hydroelectric dams, that concentrate vast amounts of energy and prepare it for distribution to other nodes. The nodes of any given coronet are arranged in multilayered hierarchies in which some nodes, the hubs, impose powerful constraints on the energetic "signals" that are relayed to other nodes. These constrained exchanges can produce complex feedback loops in which different parts of the network keep amplifying or damping the signal strength through recursive interactions.

Energy doesn't just randomly travel throughout the different parts of civilization; it cascades in highly structured patterns and organizational units. For example, the power grid is a coronet characterized by very specific flows and conversions. Fossil fuels are burned and converted into hot and pressurized steam. That steam then drives a large turbine, which rotates a giant magnet. The changing magnetic flux produces an electric field and induces a current in the coils of wire surrounding the magnet. After electricity is generated at a power plant, the resulting electric fields are carried by transmission lines to substations that regulate and redistribute the incoming voltage, and finally to local transformers that reduce the voltage further still, making it ready for end users like homes and businesses. There are roughly 50,000 substations and over 700,000 miles of transmission lines in the United States alone.[5] End users cannot simply plug in directly to power plants because the initial voltages produced there are enormous—they'd fry our houses almost instantly. For another example, consider natural gas. It's first extracted from well sites, then pumped across long-distance pipelines, then periodically pressurized at compressor stations to ensure its continuous flow, all before it finally makes its way to end users like power plants or homes and businesses. The basic reason why energy has to be refined and converted through so many different stages is because it's economically useless in its initial forms. Civilizations therefore design and build complex coronets that channel, direct, and change energy flows in highly specific patterns tailored to highly specific social needs and uses.

In a general sense, I'm going to use the term *modes* to signify the structured energy exchanges that unfold between the nodes of a particular coronet. In a simple sense, modes are the specific and refined energy signals that different parts of the coronet use to communicate with each other. Different nodes in the coronet communicate through different modes, which means that they use different kinds of signals to initiate activation and reaction sequences—in other words, they respond differently to different signals. More specifically, I'll define modes as the recursive and stratified patterns of energy flows in a given coronet. They're recursive in the sense that the output from any conversional node typically serves as an input for the next conversional node. Let's again think about the example above with fossil fuels. Coal is burned and converted to hot steam. The steam is then converted to the rotational motion of a massive turbine. The rotating turbine drives a giant electromagnet and produces electricity. The electricity is then relayed to homes and businesses where it serves as an input to devices like refrigerators and personal computers. Each step of the conversional process produces an output that becomes an input to a new conversional process. Second, modes are stratified in the sense that they're activated through a hierarchical sequence, with the energy flows becoming increasingly refined and organized along the trajectory of the sequence. As the examples with electricity and natural gas show, modal stratification is necessary to supply end users, like homes and businesses, with useful forms of energy.

Modes partially unfold across the logistical chains that link together different nodes, but it's important to emphasize upfront that they are *not* logistical chains in and of themselves. Modes are the structured energy flows that exist both *across* logistical chains and *within* specific nodes. The highly refined distribution of electricity and the distribution of natural gas are two examples of modes, as we've already seen. Another example of a modal flow is the transmission of mechanical energy across the different components of an automobile, including its pistons, shafts, flywheels, and gearboxes. As used in this context, modal patterns do *not* include the different ways of organizing labor for production and other social tasks, even though in a biophysical sense the application of human labor power certainly requires energy flows and conversions. Labor is the fundamental social source of production, and it's therefore subsumed under the political and economic conditions prevailing in different historical eras. (I'll examine its relationship to technological change and economic growth in later chapters.) Simply put, modes are the different ways of organizing energy flows to achieve specific objectives.

Energy systems are living, breathing, dynamic entities; certain modal patterns might emerge and predominate in any given time and place, but they don't remain fixed and static. People change and adapt their energy modes in response to various social and ecological challenges. This process of modal adaptation continuously unfolds throughout history. When people in England started burning more coal in the early modern period, they had to contend with the fact that it was very dirty, so they adapted modal flows to reduce the harmful effects of this excessive pollution by, for just two examples, adding bigger chimneys to houses and by redesigning furnaces used for smelting. After the Black Death killed off a large portion of the European population, there were fewer people left to feed and so those left behind increasingly started using watermills for tasks other than grinding grain, including for spinning silk and for sawing timber. Notice that modal adaptation by itself is *not* accompanied by any changes in the underlying methods of energy conversion. Whether it was for grinding grain or spinning silk, people in the Middle Ages did not have to invent completely new technologies for converting energy from some initial form to another; watermills were good enough for all of these tasks. The decisive and critical transitions in history happen precisely when modal adaptations are coupled to rapid changes in conversional methods, especially among energy-intensive technological systems, leading to a proliferation of new devices and vast expansions in an economic system's energy scale. Before fully telling that story, however, we need to have a better empirical understanding of how energy is used and measured as it circulates around the economy.

Conversion and Consumption

When most governments and organizations talk about energy use, they usually place the emphasis on *consumption* instead of conversion. Specifically, they are typically referring to a metric called *primary energy consumption*, which represents the direct use of energy sources without any prior conversions or transformations.[6] Primary consumption includes activities like burning coal at a power plant and distilling crude oil at a refinery. Primary forms of energy are not useful on their own, so they are converted and transformed into secondary forms of energy. For example, we burn coal so we can turn the resulting steam energy into electricity, and we distill crude oil so we can produce gasoline. Coal and crude oil are primary forms of energy while electricity and gasoline are considered secondary forms. The secondary sources can also be converted into other tasks and end uses, collectively known as *tertiary sources*. But there are several issues with this kind of approach to energy accounting. First and foremost, it must be emphasized that all primary energy sources are themselves the result of earlier conversions and transformations in nature, so they are not that "primary" after all. For example, the hydrocarbons of the dead plants and animals that make up petroleum are a secondary product of photosynthesis, which requires solar energy and water molecules. Just as important, primary energy metrics do not take into account the role of efficiency throughout the conversional network. Burning through 500 joules of something and getting a useful energy output of 100 joules is much better than burning through 500 joules and only getting 10 joules of useful energy.

There are two common methods of measuring primary energy: the partial-substitution method and the physical-energy content method.[7] Let's analyze them through some examples. When a power plant burns through coal, the primary energy simply equals the energy of the coal that goes up in flames. In the case of fossil fuels, then, things are pretty easy: record the amount of stuff we burn and call that primary energy. But the situation is more complicated for renewable energy sources, such as wind, solar, and hydropower, because nothing was burning while these energy sources generated electricity. Enter the two methods above. In the physical-energy content method, we simply count the electrical energy produced by these sources as primary energy, even though electricity obviously qualifies as a converted form of energy. This is the method used by the International Energy Agency to measure energy "consumption" for renewables. In the partial-substitution method, we pretend that the produced electricity came from a hypothetical thermal power plant, and then we assume some efficiency rating for this plant. For example, if the plant has a hypothetical efficiency of 20 percent, then we would multiply the electricity generated by a factor of five. In this case, the primary energy required to produce that electricity is five times larger. The company British Petroleum has adopted this partial-substitution method in its popular global energy reports.[8] The main reason why these differences matter is because they can lead to diverging estimates of energy consumption, especially for nations that rely heavily on renewables.

There's no use in arguing about which method is better. In reality, beyond the world of statistical accounting, only energetic conversions truly matter. Only energy converted to useful forms can power our economic activities. The electrical energy produced by renewables came from dynamical flows in nature, such as sunlight hitting the Earth and rivers roaring through dams. The concentration of fossil fuels at their points of processing and refinement required energetic conversions from machines and human labor, which first extracted these fuels and then transported them to a particular location. All of this happened before anything was burned and recorded on logs and charts. Thinking in terms of primary energy consumption might be useful in certain cases, but in general it obscures the energetic flows and conversions that make all economic activities possible. Recall that energy is constrained states of motion that can be exchanged among different physical systems. Also, energy can come in many different forms, such as chemical, thermal, kinetic, and potential, to name just a few. The following arguments do not even depend on any particular definition of energy; they depend on the basic fact that certain forms of energy can be converted into other forms. For example, chemical energy can be transformed into mechanical energy, which is what happens when our car engines burn through fuel and convert the resulting heat energy into the mechanical motion of the wheels. Heat and mechanical energy can also be transformed into electrical energy, as when power plants burn through coal and use the resulting steam energy to drive a generator that produces electricity.

Focusing exclusively on primary energy consumption completely ignores and marginalizes these energetic conversions, which should be the central elements of the story. This discussion brings up a major boundary problem related to our theoretical understanding of concepts like energy consumption and production. The conversion of an energy source anywhere requires the prior conversion of other sources at different locations. We think of burning coal at a power plant as "primary consumption," but the coal has to be first extracted and transported before it can be burned. That process of extraction and transportation required the conversion of other energy sources. For example, the vehicles transporting the coal may have burned gasoline or diesel on the way to their destination. It seems bizarre to exclude these activities, which facilitated the burning of coal in the first place, from the paradigm of "consumption." Of course, if we do include those activities, then we have no reason to stop there. The vehicles that transported the coal had to be built somehow, and the process of building them required other energetic inputs and conversions. The same arguments apply for any model of production. Suppose you wanted to measure the output of a car factory by the number of cars it produces annually. The production process unfolding in that factory cannot be decoupled from earlier conversions that provided the spare parts and raw materials necessary for the assembly of the cars. All acts of production and consumption are embedded within the economy's different stages of energy conversion.

These kinds of boundary problems are common to all biophysical measures that track energy flows. For example, consider the energy return on investment (EROI) pioneered by the systems ecologist Charles Hall.[9] EROI is meant to be a ratio of the useful energy extracted in an economic activity from the total energy input required to create the activity. As we mentioned before, however, energy supplied for any economic activity requires conversions from other sources and processes. This observation leads to a boundary problem about where to draw the lines on the energy supplied for a particular economic activity. In the context of existing theory, there is no good answer here. But once we begin to think about conversions as a spatially extended and interactive network, the boundary problems mostly disappear. We no longer have to analyze the energetic and economic dynamics of civilization exclusively through vague concepts like production and consumption or inputs and outputs. Instead, we just need to recognize that all acts of production and consumption are embedded within a spatially and temporally extended network of mutually dependent conversions.

What we need to understand is the extended nature of the conversional process that makes certain societies possible instead of others. There are still some boundary problems when we move the focus to conversions, but they are mostly geographic. Suppose we wanted to measure all energetic conversions for a particular nation. How should we handle activities related to international trade and commerce? If an airplane takes off from country X and lands in country Y, we would need to decide which country deserves credit for the mechanical work performed by the aircraft. Maybe we could somehow split the energy conversions between the two of them; devising a solution would not be too difficult. Although these geographic issues are very much real, they do not pose any major conceptual roadblocks. Having a geographical boundary problem about which side gets the credit for which conversion is nowhere near equivalent to the theoretical quagmire of deciding what counts as production, consumption, or "energy investment." Conversions are the ideal theoretical tool for understanding the biophysical dynamics of human economies, from which social changes emerge as chaotic adaptations.

The rate of energy conversion depends critically on the energy quality of the resources available to a group of organisms. Not all primary sources of energy are made equal. Some are more efficient than others. Some yield more mechanical work. Others produce more electricity. For example, producing one kilowatt hour of electricity in 2017 required, on average, 7,812 British thermal units (BTUs) of natural gas and 10,465 BTUs of coal.[10] By this measure, natural gas is roughly 25 percent more efficient than coal at generating the same quantity of electricity. The energy thinker Vaclav Smil identified the *power density* of an energy source as an important feature for the economic growth and development of civilization.[11] He defined power density as "energy's rate of flow per unit of surface area" of land or water.[12] This definition stands in contrast to the typical engineering definition of power density, which is energy flow through the surface area of a working device. Smil points out that for most of our history people have used low-grade, low-density energy like wood and other forms of biomass. Wood has a low power density because it takes a large surface area to facilitate the growth of huge forests, and you don't get much energy out once you burn the wood. By contrast, coal and oil usually come in compact deposits and release enormous amounts of energy once they're burned. The Industrial Revolution, for example, critically depended on a transition from low power density to high power density. Smil has argued that fossil fuels are uniquely important for modern economies because they have higher power densities than other energy sources, such as wind and solar. Larger power densities help to generate more production, which in turn leads to more commodities available for exchange and circulation around the economy. Although Smil's conception of power density is extremely useful for thinking about the design of large energy networks, it's not necessarily the most important quality for an economy's long-term sustainability.

The environmental scientist Arnulf Grubler criticized Smil's conception of power density by arguing that energy density is actually the most important constraint on economic activity.[13] Grubler asks us to consider the example of a lightning strike. It's something that delivers a huge amount of energy over a small area, and thus has a very high power density. But it only lasts a fraction of a second and could only power an electric car for less than a kilometer. In other words, a physical process that delivers large amounts of energy for a short period of time isn't very practical. What we should want first and foremost is a stable flow of useful energy. Another helpful analogy here is to imagine a water reservoir flowing through a hydroelectric dam. The amount of water in the reservoir can be thought of as a kind of energy density. The rate of flow through the dam is the power density. If a lot of water is flowing through the dam but the amount in the reservoir is very tiny, then we'll have produced a lot of energy for a short time period, yielding a large power density. However, if the amount of water in the reservoir is large, then we have access to a large pool of useful energy, regardless of how fast it's coming out. The point is that both energy density and power density are important concepts, but the availability of useful energy is what ultimately determines how much stuff we'll get done.

The concept of energy quality also includes the distribution of natural resources as well as their accessibility. An energy source can have all the power density in the world, but if it exists very far away, or becomes dispersed across a wide area, then extracting it would be an enormous challenge for any civilization. Other measures of energy quality are conceivable, but the basic point is that natural energy sources can have different uses and attributes. The only way to understand these quality differences is by looking at the conversions and transformations that follow primary consumption. Failing to take this critical but routinely ignored step makes it seem like all energy sources should be treated equally, as if they all have the same roles and functions in the economic process. The natural energy sources available for use and extraction are important not just because they constrain rates of conversion, but also because they help establish the broad parameters of technological development in any given society. People tend to develop tools and products that work with the underlying energy sources provided by their ecosystems. Technology then evolves as a social and biological mechanism for extracting and converting energy from the natural world. We can define technology as the total set of ideas and biophysical systems used or developed by humanity to rapidly extract and convert energy from the natural world. If there are deer in the forest, people invent spears so they can hunt them. If there are plants to grow and cultivate, people invent plows so they can quickly till the soil. If there are horses in the steppes, people try to ride them for transportation. If there is coal in the ground, people create machines that can rapidly burn it. If the Sun is out, people harness its rays to produce electricity. Nature provides the resources and people adapt their technological development around those resources, driven by both social and ecological imperatives along the way.

Aggregate Flow

All economies and societies extract energy from an external

environment and then dissipate it away or convert it into useful forms, such as electricity and mechanical work. When a truck moves along the highway to bring goods to the local market, it does mechanical work. When a merchant ship unloads heavy cargo containers at the port, it does mechanical work. When an airplane carries tourists from one continent to another, it does mechanical work. And when your car transports you and your family to the movie theater, it also does mechanical work. All of this work requires energy flows produced by physical gradients. Basic economic activities that are essential to our lives need energetic conversions in order to actually happen. In effect, all economies are simply byproducts of energy flows cascading through the different parts of human civilization. Economics with larger useful energy flows can generate more electricity and mechanical work, which means that they're producing more things and powering more economic activities.

Because primary energy consumption is not a great indicator of an economy's productivity, it's important to emphasize the importance of *aggregate flow*, defined as the total sum of all the useful energetic conversions that power our economic activities. In effect, aggregate flow is the total useful energy produced by an economy. It's useful energy that allows civilizations to make the goods and services consumed by people. A related quantity of interest is the *aggregate flow rate*, or AFR, which measures the aggregate flow per unit of time. The AFR per capita can serve as a powerful metric of economic growth or decline in biophysical terms. Economic growth is an annual gain in the AFR per capita. Meanwhile, economic decline is an annual drop in the AFR per capita.

Unlike GDP, aggregate flow can actually be aggregated because energy is measured through a stable unit, the joule. The AFR is a reliable measure of economic scale and growth because it directly measures how much useful energy people are able to produce through their technological systems. Furthermore, the evolution of the AFR per capita corresponds strongly to our intuitions about "rich" and "poor" societies. Wealthy societies can convert more energy per person than the poor ones. Using aggregate flow would remove all intellectual and theoretical doubts about aggregation in economics. The main disadvantage relative to the current metrics is that the AFR is more difficult to measure than GDP. But it's not that much more difficult, and statisticians could always use models and assumptions to fill in gaps about energy conversions in order to arrive at a final number. Moreover, we're not limited to only the AFR. We can measure the physical size of an economy in many ways. We can measure the total weight of everything that an economy produced in a given year, we can measure an economy's total energy consumption, and many other quantities as well. All of these quantities would be far better measures of aggregate productivity than financial aggregates. If we measure economic size through energy consumption, for example, then China would be the world's largest economy by far.

There are many ways to expand the AFR of an economy. An obvious way is to improve the aggregate efficiency of the economy while holding total energy consumption constant, or to improve aggregate efficiency so much that total energy consumption can even be allowed to decline and the AFR would still go up. This latter option would be the ideal scenario, but it's extremely difficult to pull off in the real world, as we'll shortly see. Another way is to increase total consumption while holding aggregate efficiency roughly constant. This is the more typical approach used by capitalist economies in the short run. Of course, it's also possible to increase the AFR by letting aggregate efficiency decline as long as the rise in total energy use is large enough to compensate. This latter scenario was precisely the one that unfolded for the initial phases of industrialization in England during the eighteenth century and the early nineteenth century. If we wanted the AFR to decline, we would just take the inverse of all these cases.

We now have a more holistic view of how society uses energy through different nodes embedded in a complex network. We live in an age of great ecological uncertainty caused by the energy-intensive demands of capitalism. Having a network as a metaphor makes it easier to understand the biophysical changes that should occur in a new economic system: the way energy is produced, the way it's distributed, and the way it's consumed. The future prosperity of civilization depends on a successful transition to a different kind of conversional network, one that burns far fewer fossil fuels, as well as one that is managed and controlled by different social and economic relations. But if we're going to carry out this transition, we first need to carefully think about how to do it, because a transition on that scale will be difficult, complex, and expensive. That's the main reason why we need more democratic and collective control over the evolution of our economies and their associated conversional networks.

CHAPTER 6

The Ecodynamic Synthesis

Human economies and the natural world are an integrated system. The two evolve interdependently, meaning that changes in one domain can reverberate to the other, and vice versa. This is the fundamental assumption of the ecodynamic synthesis. Natural ecosystems supply our economies with the energy flows necessary to generate useful work, and our economies use that work, along with the energy wasted and dissipated, to alter and reconfigure the natural environment. There's an ongoing dynamic feedback loop between the two systems, which are always interacting together in highly chaotic and unpredictable ways. Recall that the central tenets of the ecodynamic synthesis can be captured through a flow-cycle model involving nature and society, in which nature is the flowing river and society is the rotating watermill. The flows are collections of energy and materials that economies absorb from the natural world. Some portions of these flows are used to drive economic cycles, which are the recurring patterns of production, distribution, and consumption. But the flows that cascade through civilization are for the most part lost to the global ecosphere, where they amplify preexisting natural instabilities that can become highly disruptive to the prevailing social order as time goes on.

The ecological realm represents the dominant causal order of economics and human societies. Economies are complex biophysical systems that interact with the wider natural world, and none can be fully examined apart from their underlying material conditions. Formally, we define an economy as an open and dissipative thermodynamic system in which people and the technologies they control exchange energy with an external environment. Economies function like dynamical systems because they operate far from thermodynamic equilibrium, meaning that a net flow of energy exists between our economic systems and the wider ecological order. Studying and understanding the nature of these material and energetic flows is the chief task of economics as a scientific discipline. Because economies are dynamical systems, they necessarily experience energy losses through waste and dissipation. Broadly, all economies convert the energy they consume either into electricity and mechanical work, along with the products derived from that work, or into various forms of waste and dissipation. We can define the aggregate efficiency of an economy as the share of all the energy consumed that goes into the conversion of mechanical work and electrical energy. If two economies have different aggregate efficiencies, then the one with the higher efficiency will produce more useful energy. The remaining energy balance, whatever did not get converted, is lost through various methods of waste and dissipation. These dissipative losses can no longer provide useful work for the economic system, but they are still important to its future evolution. Substantial waste and dissipative losses have the capacity to profoundly change the distribution of energy in the local ecozones and the global ecosphere that the economy needs to function.

All human societies are embedded in the ecosphere and are reliant on it for their survival. Marx understood that human existence and human labor are utterly cointegrated with the natural world:

> Labor is, first of all, a process between man and nature, a process by which man, through his own actions, mediates, regulates,

and controls the metabolism between himself and nature. He confronts the materials of nature as a force of nature. He sets in motion the natural forces which belong to his own body, his arms, legs, head and hands, in order to appropriate the materials of nature in a form adapted to his own needs. Through this movement he acts upon external nature and changes it, and in this way, he simultaneously changes his own nature.[1]

The relations between society and nature are historically contingent, and that's especially true in the era of modern capitalism, where the internal dramas and conflicts of human society are often externalized and offloaded to nature itself. In other words, the internal class dynamics of capitalism are only possible because of the system's reliance on limitless extractivism. The growth of capitalism has been accompanied by the degeneration of the natural world. This tense dialectical relationship has been explored by many ecological thinkers and scholars over the years. Marx extensively analyzed the ecological impacts of capitalist production and realized that capitalism leads to ecological degradation because it does not replace what it extracts from nature.

Capitalism conceives of the natural world as a means to an end: profit and class domination. Ecological stability is not good for corporate accumulation, so nature becomes a bottomless pit for capitalist extraction and expropriation, with no worry of regenerating or replacing what has been seized and squandered. In particular, Marx focused on the degradation of soil quality in England under capitalist agriculture, which fed an imperialist drive to find fertilizers worldwide that culminated in the extraction of guano from South America using forced labor. Marx's work has inspired the modern framework of *metabolic rift theory*, which has been thoroughly developed and analyzed by thinkers like John Bellamy Foster, Brett Clark, Paul Burkett, and Kohei Saito, among many others. In their 2020 work, *The Robbery of Nature*, Foster and Clark defined the metabolic rift as the "active estrangement" of the social metabolism from the natural world.[2] The concept of metabolic rift posits that the expansion of capitalism is undermining the ecological basis for the survival of human civilization. Foster and Clark write:

> Like any complex, dynamic system, capitalism has both an inner force that propels it and objective conditions outside itself that set its boundaries, the relations to which are forever changing. The inner dynamic of the system is governed by the process of *exploitation* of labor power, under the guise of equal exchange, while its primary relation to its external environment is one of *expropriation* [appropriation with no equivalence].[3]

Other thinkers have also analyzed the complex dialectical relationship between modern capitalist society and the natural world. In *Silent Spring*, Rachel Carson argued that the economic activities of human civilization can have profound effects on the natural world. She demonstrated these effects by examining the widespread use of pesticides, which contaminated local environments and harmed biodiversity. Her works had important social and political impact, inspiring nations around the world to ban harmful pesticides and quickly launching a wide array of environmental movements. Other major contributions to ecological theory include *The Entropy Law and the Economic Process* from the mathematician Nicholas Georgescu-Roegen, who argued that our economic activities are extensions of our biophysical constraints and realities. He viewed economics as a thermodynamic process that leads to greater entropic production, providing one of the first major attempts to marry fundamental physics with economic theory. Although he was rightly criticized for incorrectly presenting some issues related to thermodynamics, his work nevertheless inspired a whole new generation of thinkers to examine the physical relationships between our economic systems and the natural world in which they function. The Canadian ecologist William Rees, in works such as *Our Ecological Footprint*, aimed to quantify the burdens that our prevailing resource consumption strategies under modern capitalism are imposing on the natural world. Rees emphasized the need to pursue a sustainable developmental path consistent with long-term ecological constraints. From a purely ecological standpoint, the scientist Howard Odum made the most important theoretical contributions to the study of ecology by emphasizing the perspective of general systems theory. He viewed the economic and ecological worlds as integrated systems interacting through complex energy flows. On the empirical side of things, a variety of prominent thinkers, including Robert Ayres, Benjamin Warr, and Carey King, developed some of the theoretical machinery for understanding the material and biophysical foundations of economics.

Ecology and Humanity

In the flow-cycle model of the ecodynamic synthesis, natural energy flows constrain and facilitate economic cycles by providing logistical channels and resource bases for extraction. However, the economic cycles of civilization can also impact ecological flows through various amplifier effects and other nonlinear perturbations, because the waste that we throw back out to nature often serves as an energy reservoir for other biotic and abiotic systems. Although resource availability is critical for economic activity, since we can't have finished manufactured products without the raw ingredients from nature, it's not necessarily the most important effect that ecology has on the dynamics of human civilization. Neoclassical theories have identified the central problem of all economics as being the scarcity of resources. The basic argument is that we have infinite wants but limited resources, and because we cannot have everything we want, we must necessarily devise a system to distribute goods and resources. Enter the market economy with its prices and wages, which are supposed to act like gatekeepers to the warehouse of economic nirvana. There's a kernel of inadvertent truth behind this story. Natural limits certainly impose absolute scarcities that are impossible to overcome.

THE ECODYNAMIC SYNTHESIS

There's only so much uranium in the Solar System, for example. And even if we synthesize certain substances by using other substances, the total amount we can produce will still be limited by the availability of the raw materials going into the production process. We can't beat energy conservation.

Although natural constraints on supply are important, most of the economic scarcities that rule our lives are actually social and artificial, not natural. As we've already seen, supply and demand are not magical forces drifting through the air; they're contrived realities established by an interactive social environment involving governments, corporations, institutions, and classes. Supply and demand cycles are social constructs designed to answer a basic question: Who gets what? Those with social and institutional power decide how they want to distribute money, labor, and resources, and those without such power must navigate the resulting constraints and roadblocks that have been thrown in front of them, or they can challenge the system and remove some of the roadblocks, if not all of them. Especially under capitalism, scarcities are usually artificial; they're created by certain classes and power blocs to boost their rate of financial accumulation. Artificial scarcity is an important *social* reality that torments the lives of billions around the world, but scarcity as a natural limiting factor in economic activity is not as fundamental as we might like to think. In that case, what is?

Let's begin to answer this question by remembering that human economies are dynamical systems powered by energy flows, and their successful operation requires the presence of *stability* in the face of an uncertain environment. If ecological instabilities make it difficult for an economy to access existing energy stocks, then that economy is susceptible to collapse even though in theory plenty of energy remains available for consumption. In 2020, the coronavirus pandemic painfully revealed this fundamental truth once again. The global economy experienced the worst cataclysm since the Second World War not because economic and natural resources suddenly vanished, but because chaotic feedback loops between nature and society destabilized societies and governments worldwide, making it difficult for many people to carry on with their previous way of life. Thus, the central problem of economics is not scarcity, but *stability in the flow of goods and resources*, and especially the stability of the ecozones that act as an economy's primary energy reservoir. It's precisely this stability that's threatened by the energy-intensive nature of capitalism, as the massive energy losses dumped into the biosphere risk triggering ecological tipping points and thresholds that could overwhelm modern civilization's ability to function normally. The primary goal of any economic system should therefore be to ensure stability and sustainability in the face of nature's external perturbations, which have always played a dominant role in the development of human history. Before going further, it's worth having a general picture in mind about what stability means at a theoretical and empirical level. We cannot pursue stability as a strategy unless we know what we're trying to stabilize, and why it's worth stabilizing in the first place. In this book, stability is understood as something like a *dynamic equilibrium*, an acceptable range of aggregate energy use for human civilization that allows it to function without transgressing critical planetary boundaries. People are complex, to say nothing of entire societies. No civilization would be able to maintain a constant rate of energy use at all times, which is why viewing stability as a constrained dynamic equilibrium offers civilization more balance and flexibility as it tries to coexist with the natural world.

Although achieving stability is the central goal of any economy, change happens all the time in our ecological and economic systems, whether we like it or not. The most important features of human history result from economic transitions. We shall think of an economic transition as being characterized by a substantial change in the AFR, whether upward or downward, from some collection of ecozones. The transition is complete once this new conversion rate stabilizes or hovers near some long-term equilibrium value. During the transitional period, the economy finds itself in a critical state characterized by major technological changes, class struggles, social upheavals, and political quarrels. Initially, large increases to the rate of conversion usually require some kind of ecological primer that causes social instability, thus giving people incentives and opportunities to produce and consume more energy. Think of the last Ice Age coming to an end, or historically more recent events like the Black Death, which I'll cover in more detail shortly. Natural ecosystems provide energetic constraints on any given economic system, but those constraints can be overcome, and have been at various points in history, if some powerful ecological primers converge at the right time. This usually means that ecological changes combine to yield more natural resources that people can exploit, but it could also mean that ecological changes force people to migrate and to explore other areas, leading to the discovery of new lands and economic opportunities. These ecological changes and disruptions can be rapid or they can be gradual, but in the end they must be strong enough to establish new energy flows that either reinforce or disrupt the preexisting ones.

Economies with high AFR values require more energy-intensive resources than those with low AFRs. Modern capitalism is addicted to fossil fuels in large part because fossil fuels have substantial energy densities; enormous amounts of concentrated energy flows can be produced when burning coal, oil, and natural gas. And the consequences of extracting these resources and deploying the corresponding technologies can be profound. For example, someone picking cotton with her bare hands will not collect as much as someone going around with a tractor doing the same thing. And weavers in ancient times could not match the massive textile production of modern factories. The dominant resources and technologies in any economic system constrain or influence several important features of that system, including the distribution of wealth and the resulting class-based social hierarchies. Likewise, those social features constrain and influence the evolution of technological change and resource extraction. All of these factors collectively converge through highly dynamic and nonlinear processes to generate globally coherent features in their underlying coronets.

A Tale of Two Islands and Three Nations

Fundamental lessons about the interplay between ecology and economics can be found among societies throughout history. To explore some of the critical ideas presented so far, and to understand the material relationships between human societies and the natural world more concretely, we'll look at some instructive historical examples of societies that faced long-term ecological challenges. These examples underline the physical constraints that shape and produce the ecological evolution of human civilization. No two economies are exactly the same because no two economies developed from exactly the same physical and ecological conditions. Economies may differ in size, in the way they distribute resources, and in how much energy they consume from their local ecozones. Let's begin with the Polynesians, a group of settlers and explorers who became famous for their epic voyages throughout the Pacific. They discovered Hawaii, Tahiti, New Zealand, and hundreds of other islands. Because their core territories were so far away from the rest of the world, they offer an excellent opportunity to study isolated economic systems operating in local environments. The Polynesians struggled with economic growth and survival on islands that had few, or very limited, natural resources. They attempted to compensate for these deficiencies through trade and conquest, with varying levels of success.

The Polynesians' experiences offer some profound lessons and warnings for the rest of humanity. The tiny island of Tikopia, located in the Pacific, was first inhabited back in ancient times and remains one of the canonical examples of an economic transition induced by ecological factors. Tikopia can hardly support a population much bigger than a thousand people. The island is so small, and the resources so few, that the Polynesian settlers carefully controlled the number of children they had and the amount of food they could consume.[4] When they first discovered Tikopia, the brave explorers brought along not just their outrigger canoes, but also their personal belongings and domesticated animals. As the centuries went on, the islanders eventually realized that one of their most prized possessions, their pigs, had become so numerous that they were draining the island's food supply. In response, the islanders collectively decided to eliminate their pigs from the island, an act tantamount to abandoning a major symbol of power and wealth. They committed this desperate act because they realized their island was a fragile ecosystem that could only support so much life. Something had to be sacrificed, or else Tikopian society would devolve into warfare over the remaining scraps and run the risk of completely collapsing. In addition, the Tikopians gradually phased out intensive agricultural practices that harmed the soil and the island. The stable modal patterns of the Tikopians made them one of the most successful societies in the Polynesian world, allowing them to thrive all the way to present times, long after groups in other Polynesian islands had already collapsed.

Other Polynesian societies were not as fortunate as the Tikopians. Easter Island is often held up as the canonical example of a place where communities pursued the wrong modal adaptations in relation to their environment. The traditional story goes that the Rapa Nui people who populated the island centuries ago managed to quickly dwindle critical resources, such as trees and wildlife, through overconsumption. The resulting resource scarcities and extreme levels of deforestation produced a breakdown in the social order of the Rapa Nui, leading to wars and famines that ravaged the population of the island.[5] Other scholars, however, have challenged the traditional narrative.[6] They argue that the Rapa Nui were actually ecological in the resource strategies they adopted, and they blame the rats tagging along in the canoes for the widespread deforestation of the island. Contrary to the traditional version and its emphasis on conflict among the islanders, the revised version claims that contact with Europeans was the ultimate development that sealed the fate of the Rapa Nui. The diseases introduced by European explorers increased the death rate among the Rapa Nui and sparked enormous tensions, which apparently culminated in major conflicts on the island during the nineteenth century, a far later date for the collapse of Rapa Nui society than that suggested by the traditional story. But whichever version is correct, our central argument does not change: isolated societies and ecological environments face unique and difficult challenges. These can include everything from the loss of local biodiversity in response to the introduction of new species to the inability to replace shortfalls in local resources by finding trade partners elsewhere. Isolated economies can certainly overcome these challenges for long periods of time. However, doing so requires making sacrifices and careful planning. Virtually all Polynesian societies struggled with the same ecological issues as Tikopia and Easter Island, and their experiences offer important lessons that echo through the ages.

Polynesian societies are great case studies in the importance of ecology because of their general isolation from the rest of the world, but ecological factors have played a dominant role in the long arc of historical development among all advanced civilizations. Three examples of these advanced civilizations and the ecological and geographical situations they faced at specific points in time are China under the Song dynasty, Japan under the Tokugawa Shogunate, and England in the early modern period. The Song dynasty was a period in Chinese history lasting from 960 to 1279. Song China is the greatest example of a premodern society that almost pulled off a major industrial revolution. Aided along by a highly stable climate, Song China experienced numerous bursts of rapid technological development and economic innovations, including the first documented use of paper money in human history and the invention of movable type printing. The surge in iron production was one of the era's defining features. In 999, Chinese tax officials calculated total iron output in Song China to be roughly 32,500 tons, but by 1078 that number had risen to an astonishing 125,000 tons, a four-fold increase in annual output.[7] The Chinese initially produced iron for agricultural tools and later started focusing more on weapons. In addition to gunpowder, they also produced crossbows and arrows. At its height, Song China may have manufactured 16 million iron arrowheads every year. It's likely that Song China had "the highest iron output per capita in the world until 1750," when England started breaking records as part of its industrialization wave.[8] Roughly a million Chinese households may have been involved in the production of iron by the eleventh century.[9] Historian Arnold Pacey summed up Song China's technological development:

> In 1100, China was undoubtedly the most technically "advanced" region in the world, particularly with regard to the use of coke in iron smelting, canal transport, and farm implements. Bridge design and textile machinery had also been developing rapidly. In all these fields, there were techniques in use in eleventh-century China which had no parallel in Europe until around 1700.[10]

The rise of markets and urbanization facilitated the development of domestic and external trading networks. Almost a thousand years before the rise of Western capitalism, China had created the largest and most dynamic market in the world. The expansion of cities in Song China also contributed to higher rates of deforestation, particularly in the north of the country. In what's going to become a familiar story, Chinese households and businesses started to replace timber with coal for their heating needs, leading to booming coal markets and industries along the way. Coal became especially popular as the heating fuel for iron smelting, and coke-burning blast furnaces started to proliferate in northern China. Scholars have estimated that at least 140,000 tons of coal were used by the iron industries every year in the border regions of Henan and Hebei, two provinces in east-central China.[11] But despite these rapid advances in the adoption of coal and iron, Song China never managed to fully industrialize quite like the English would a few centuries later. Why is that? Why wasn't China the first country to complete its own industrial revolution? Why wasn't China the first country to rule the world? There are several reasons for its unique historical trajectory.

The first is that the sheer geographical scale of China made it difficult and expensive to transport coal over long distances. As a result, the "industrial and household use of coal remained highly localized," with coal prices sometimes reaching five times higher at markets along the Yellow River compared to the price at inland mines that were only 50 kilometers away.[12] These extreme price differentials were a powerful barrier to increasing coal consumption in China. In a related reason, coal deposits were vastly concentrated in northern China, with southern China holding limited stocks. That lopsided geographical distribution made a difficult transportation problem even worse, as the southern regions would be entirely dependent on the northern regions for coal, a reliance that would cause prices to skyrocket and that would therefore make coal unaffordable. A second reason for China's failure to industrialize is that the Chinese never had an acute labor shortage. As a result, they had weak incentives to invent labor-saving machinery like steam engines. Although the Chinese also had many problems extracting coal from underground mines, such as flooding and poor ventilation, they relied much more on human labor to overcome these challenges. Because labor was relatively cheap, producers and the Chinese government never made a huge financial investment for the large-scale development of machinery and industry.

Finally, we can't ignore the role of geopolitics. China was historically surrounded by semi-nomadic groups of people that periodically invaded their wealthier neighbor. The Song dynasty did not manage to escape this violent historical pattern. In 1127, the Song capital of Kaifeng fell to an invasion from the Jurchen peoples, who lived in the northern periphery of China, an area commonly called Manchuria in modern times. The Jurchen conquered most of northern China, establishing a new state called the Jin dynasty, while the Song aristocracy fled to the south and created a new power base there. Because the Song lost control over the lucrative coal deposits in the northern regions, it became practically impossible to carry out a successful industrialization strategy purely with the limited resources of southern China. Had the Song been given another century of peaceful internal development, it's extremely likely they would have become technologically advanced enough to permanently conquer or subdue their seminomadic neighbors, including the Mongols, and the borders of Asia today would be radically different. Things didn't turn out that way, however. By 1279, it was the Mongols who actually conquered the Song, establishing the Yuan dynasty in the process. With that political denouement, the moment for full-blown industrialization had passed, and China wouldn't seize the global economic spotlight again until modern times.

In the seventeenth century, something remarkably similar happened to the people of England and Japan: they experienced a deforestation crisis, with the one in Japan slightly more severe. This unusual shortage of timber prompted major changes in both societies. England eventually resolved its ecological problem by burning coal instead of wood, launching the world's first industrial revolution along the way. The most important reason why the English successfully made this substitution is because they could. England was "blessed" with vast amounts of coal deposits *that were easily accessible*, meaning they were close to major towns and rivers. On the other hand, Japan did not have the same natural fortunes, and could not carry out a similar kind of internal substitution. Japan could only solve its ecological problem by *reorganizing* its economy, which meant implementing new social and political rules to reduce energy use, to limit the extraction of natural resources, and to establish far stricter policies related to forest management, such as village quotas on the amount of wood that could be harvested annually. Through these policies, Japanese authorities restricted the country's economic and demographic growth for over a century, but they also saved the country's forests and prevented an otherwise inevitable ecological collapse.

Japan is an archipelago that contains thousands of islands, the largest of which is Honshu. The first groups of people arrived in these islands thousands of years ago, but they were later joined by waves of new migrants that contributed to the emergence of a unique and relatively isolated culture. As the population of Japan increased after the adoption of agriculture, trade and commerce with other states and cultures throughout East Asia became more prevalent. By the Middle Ages, Japanese society featured a strong feudal hierarchy in which aristocratic lords, the *daimyo*, ruled over the peasants and villagers that constituted the bulk of the population. The limited supply of arable land and other critical natural resources meant that warfare was a regular staple of Japanese history. The wars required large quantities of timber for the completion of ships, forts, and weapons, especially during the violent Sengoku period in the sixteenth century. Massive construction booms also used large amounts of timber, a trend accelerated by higher levels of trade and commerce as a result of the peaceful period established by the Tokugawa shogunate in the seventeenth century. These energy-intensive economic activities caused Japan to reach a state of ecological crisis at the end of the seventeenth century. Entire forests on the major islands had vanished from excessive logging, causing severe soil erosion and disrupting the basic structure of the Japanese economy.

Hoping to stave off economic disaster, the Tokugawa shogunate responded by imposing higher levels of centralized control over the economy. This comprehensive response, a combination of top-down control and local control, included regulations on forest management, transportation controls, and constraints on timber consumption.[13] The Shogun authorized magistrates to impose rules related to forest management on districts throughout Japan. The Shogun also specified the amount of wood that people in different social classes could use in the construction of their homes. For example, a leader presiding over several villages could use up to 30 *ken*, with a ken being a six-foot-long beam of wood.[14] A taxable peasant could use six ken and a fisherman could only use four.

The Tokugawa also encouraged fishing in open waters and trading with other indigenous groups in the archipelago as a way of reducing reliance on intensive agriculture. The Japanese also created special plantations and quota systems that carefully regulated the amount of wood that people could cut. In addition, guards along roads and rivers checked people who were transporting wood to make sure they followed regulations. Japan also benefited from ecological factors, like high rates of rainfall that allowed forests to quickly regrow and seafood to proliferate. Politically, the centralization of power under the Tokugawa also allowed Japan to implement and *enforce* ecological regulations.

These policies collectively resulted in a stable rate of energy use and little population growth from 1700 to the early nineteenth century. From 1721 to 1828, Japan's population barely budged, increasing from 26.1 million to 27.2 million.[15] But the Tokugawa policies also did their job; they led to the reforestation of Japan by the end of the nineteenth century, avoiding what would otherwise have been a catastrophic economic collapse. Japanese civilization survived. What happened in Japan remains one of the most impressive cases of large-scale ecological adaptation in human history. In the nineteenth century, Japan began trading and interacting with the West. It imported commodities, raw materials, and technical expertise from various European countries. As a result, Japan settled on a path of rapid industrialization that made it one of the world's great powers by the beginning of the twentieth century. Today, Japan continues to face enormous resource challenges exacerbated by its energy-intensive economy. As its rate of energy conversion stabilized after reaching incredible heights in the twentieth century, generating more economic growth increasingly became a major struggle and the country's population fell substantially. Japan's ultimate fate is now linked to the general developments of this marvelous and unpredictable planet.

Japan's enormous changes under the Tokugawa reveal that successful ecological transformation on rapid timescales requires both some level of top-down political and economic control as well as more local forms of ecological and community management. These two sets of social institutions can serve mutually reinforcing and overlapping constraints; they don't have to stand in conflict or in opposition to each other, which is how their relationship is often framed. Japan is also a reminder that ecology is not necessarily destiny, at least over short periods of time. Even though the Japanese archipelago had limited natural resources, the Japanese people still managed to industrialize their economy and to reach an incredibly high standard of living. Many people see civilizations like Japan as proof that ecological constraints are not important, or perhaps not decisive for economic growth and development. But there is no contradiction in recognizing that ecology plays the fundamental role in the long-term development of civilization and also in recognizing that some civilizations can export their modal patterns to others, which is what happened between Japan and the West. Trade and commerce do not disprove the centrality of ecology; they just remind us that we have to shift the scale of analysis. When multiple civilizations trade and communicate with each other, the ecological order must expand to include all of their interactions. If we really want to understand how and what Japan learned from the West, we also need to consider the ecological order that sustained, pressured, and forged the basic contours of Western civilization. Then we can gain a broader picture on the economic history of the West. We can also compare that path, as historians like Jared Diamond and Ian Morris have done, to the trajectories followed by other regions and civilizations around the world.

The subtitle above promised a tale about *three* nations. But before properly telling the story of how the English industrialized, it is necessary to have a much better grasp on the mutual and convergent interactions between technology, finance, and energy, in addition to understanding how social classes and hierarchies mediate the pace and structure of these interactions. On a philosophical note, understanding the central importance of ecology for human civilization does not diminish the power of human action. We do not stand above, below, or apart from nature. We belong to nature, as emphasized in prior chapters. But our capacity to act in the world is limited, and here is where ecology reveals its guiding hand. Ecology does not determine every feature of society, but it does determine the absolute limits and boundaries that restrict society. Depending on the shifting circumstances of the wider ecosphere, it may also determine whether society in any normal sense is even possible in the first place. These limits and boundaries then unleash complex and chaotic interactions that separate one social trajectory from another. Humans are still very much the most entertaining actors in this mess we call history, but we are not the only ones.

CHAPTER 7

Technological Dynamics of Growth and Stability

How does technological change produce economic growth? And how does this growth affect the broader ecosphere in which it's embedded? These are the central questions to which we now turn. Recall that, in the ecodynamic synthesis, growth means the expansion and diversification of useful energy forms, where useful energy can be anything from mechanical work to electricity. The AFR is the total measure of the energy converted by a particular economy in a given year, so growth just means an increase in the AFR. We can think of technology as the set of all biophysical systems, productive methods, and intellectual abilities that allow human beings to convert energy from a preexisting form to another form. Nature doesn't hand us pottery, tables, and cars. These things must be built and manufactured from raw materials, and that process involves everything from labor power to machines and vehicles. For example, coal power plants burn coal to produce electricity, so we'd say that they've converted coal into electricity. For another case, natural cotton fibers first need to be carded and spun so they can be straightened and strengthened, then the resulting threads are woven into finished articles of clothing.

The main focus in this chapter is to analyze the intersection between ecological dynamics and technological change. It would be a mistake to see technological change as being exclusively a *socially* immanent phenomenon, which in this context means a process that's happening within human society. Societies interact with the natural world in highly complex ways, and those interactions also affect the pace and direction of technological development. The main point is that technological change has no inherent goal; technology is an immanent and emergent feature of the nexus between the social and ecological worlds. Technological changes are shaped and constrained by political conflicts, class struggles, as well as ecological primers that induce different modal adaptations. Nor is technology merely a fixed set of products at a given point in time; as part of the conversional process, it's also a method of adaptive engineering in the face of complex problems and challenges. Nevertheless, technological change is a process embedded within society, hence it both affects and depends on social, political, and cultural relations, such as trade, commerce, and money. To sustain the energy-intensive and multidimensional conversional stages of their broader coronets, people usually establish new institutions and organizational structures. Labor relations change and new social hierarchies are erected. If a new conversional process requires more labor-intensive work, then new forms of management and administration will also be required to control and direct the available labor power. That process can lead to the verticalization of hierarchy, with new layers of management needed to control the conversional process. Changes in the conversional process necessitate changes in specialization and the division of labor. If a new conversional process requires greater energy flows and extends longer across time and space relative to an earlier process, it can lead to new modal patterns, social hierarchies, and organizational structures. That's because a complex and extended conversional process that requires multiple stages of transformation needs different people who can master or supervise the different stages.

The concept of "technological innovation" is ubiquitous in modern society, but it's also more controversial and open to interpretation than one might normally think, and there's no universal agreement on its empirical definition. Some scholars have tried to measure innovation at the national level by analyzing the number of patents produced by a particular country, the idea being that more approved patents indicate the invention of more valuable methods and technologies. But this metric certainly has major flaws, especially for the United States. One study estimated that almost 30 percent of all American patents would be found invalid if challenged in court.[1] Indeed, for many years in the 2010s, roughly half of all software patents taken to court were struck down and invalidated. The United States Patent and Trademark Office has a notorious history of approving ridiculous patents, including for products like urinal headrests and odors that supposedly cure male impotence. R&D funding is another metric that has been used to study technological innovation. Nevertheless, it too has many problems. Spending lots of money pursuing a technological objective doesn't always lead to innovation. Just look at the money that's been wasted on nuclear fusion without any concrete results. Moreover, corporations often play around with the meaning of R&D. Pharmaceutical companies spend the vast majority of their "research" funds on developing variants of preexisting drugs so that they can extend patent protections that would otherwise expire, posing a serious threat to profits.[2] Independent studies show that these variants are hardly ever more effective than the older versions of the drugs, although the industry funds many of its own flawed studies that eventually make their way to the FDA. Big Pharma then turns around and argues that they deserve their ridiculous profits because they spend so much on innovation. The truth is that a lot of what gets labeled as "R&D

funding" is going toward projects that have no direct impact on technological change. And finally, there's a common trope that deserves to be mentioned in the neoclassical theory of growth: total factor productivity (TFP). This is supposed to be the share of economic growth that cannot be explained by labor and capital inputs. For all practical purposes, it's understood as the share of economic growth that's explained by technological progress and innovation. In reality, it's just an algebraic factor that appears in neoclassical aggregate "production" functions, which reveal nothing about production because they're little more than accounting identities that are true by definition.[3] TFP is perhaps the most egregious intellectual embarrassment of modern macroeconomics. It implies absolutely nothing about technological innovation and should therefore never come up again in any serious discussion on the subject.

Theories of Technological Change

The subject of technological innovation has aroused no shortage of theories. Technology has been analyzed as everything from a kind of knowledge to a particular set of empirical examples. The historian Alex Roland defined technology as the systematic and goal-oriented manipulation of the natural world, writing that technology "has four components: materials, technique, power, and tools or machines. Thus, technology is the process of applying *power* by some *technique* through the medium of some *tool* or *machine* to alter some *material* in some useful way."[4] It's quite a mechanistic conception of technology as tools and techniques that are applied to change some underlying material. A historical example of this conception would be the adoption of the heavy plow in medieval Europe, which caused a large increase in agricultural productivity and population growth, according to historian Lynn White and other scholars.[5] Another famous historian of technological development, Thomas Hughes, claimed that technology is "the effort to organize the world for problem solving so that goods and services can be invented, developed, produced, and used."[6] This kind of view hints at the more social, economic, and even political aspects of technological development. Thinking along these lines, the theorist Langdon Winner suggested that technological systems are the embodiment of social relations, even political ones. He famously claimed that technologies are forms of life and that "we do not use technologies so much as live them. One begins to think differently about tools when one notices that the tools include persons as functioning parts."[7] Socialist thinkers have also contributed much to our understanding of how technology relates to labor and class. In volume 1 of *Capital*, Marx famously analyzed the role of technology in the production process, arguing that capitalists use automation as a way of replacing or controlling their labor force, thereby reducing labor costs and temporarily boosting profits.

Theorists of complex systems view the process of technological change as a set of entangled interactions between various components, agents, social groups, and institutions. Thomas Hughes introduced the concept of a *reverse salient*, which is a component of a technological system that functions abnormally and therefore prevents the full development of the system. A major historical example of a reverse salient were the boilers used in the first steam engines. Because early boilers relied on low-quality metals and poor construction techniques, they could not fully support the operation of high-pressure steam engines, largely because that high-pressure gas would find its way through small cracks and crevices, causing massive explosions. It wasn't until the early nineteenth century that industrial ironworks had improved enough to allow for the construction of quasi-stable boilers, thus partially facilitating the diffusion of high-pressure steam power. One of the more popular theories on technological development is path dependence. According to this view, the development of technology follows a clear historical pattern, implying that the future depends on the past. History matters, in other words. The innovation of future tools, products, and methods follows from the development of previous tools, products, and methods; there's a "path" from one to the other, so to speak. A classic though controversial example of path dependence is the QWERTY keyboard, which remains in use despite the higher efficiency of alternative keyboards.

The basic point is that technological change is generally gradual, additive, and cumulative, not sudden and revolutionary. Rarely is something completely new invented; existing technological components are instead combined and integrated into hybrid devices and designs, which can be considered new in the sense that they perform novel functions or solve previously intractable problems. Consider the flywheel, one of the simplest yet most powerful technologies humans have ever created. A flywheel is a rotating disk that's used to conserve rotational kinetic energy. Flywheels are basically used in devices that rotate to help modulate their speed of rotation. Over time, flywheels have been attached to pottery wheels, watermills, steam engines, and internal combustion engines. Your car engine has one right now. Whenever you start your engine and get going on your drive, the pistons getting pushed by the combustion strokes are turning a crankshaft mechanism that is attached to a flywheel. The metallic flywheel helps to conserve the rotational energy of the crankshaft even when you're no longer pressing on the gas pedal, which prevents the engine from stalling and gives you a smooth drive. The crankshaft itself is another great example, as it has existed at least since the Middle Ages and has been widely used to convert linear reciprocating motion into rotational motion for multiple mechanical devices and components.

Other popular theories hold that technological change is *evolutionary*; just as organisms that are well adapted to their environment will survive and produce offspring, technological systems that are selected by their wider social environment will survive while others disappear. Indeed, theories of technology have been richly shaped and influenced by the work of numerous biologists. Humberto Maturana and Francisco Valero introduced the concept of *autopoiesis* to describe systems that are capable of self-reproducing their elements and parts, such as cells.[8] By contrast, *allopoietic* systems use energy and resources to produce other things, like a factory that produces cars. An entire economy can be seen as an autopoietic system consisting of allopoietic components. In an effort to better describe how life works, the biologist Stuart Kauffman introduced the idea of gradient-driven *autocatalytic sets*.[9] An autocatalytic set is a collection of molecules that undergo chemical reactions through a processing sequence driven by gradients with the external environment. In this context, the term *processing sequence* means a sequentially activated set of biochemical reactions that produce special molecules that further catalyze, or speed up, at least one of the reactions in the set. The external gradient keeps driving this self-organized feedback loop by providing the necessary energy flows, without which the autocatalytic sets could not be activated. The broader environment is absolutely critical in shaping the physical functions and processes of all autocatalytic sets. The flow-cycle model of the ecodynamic synthesis is highly analogous, as natural energy flows are in effect biophysical gradients that facilitate the dynamic evolution of our conversional networks, which operate through globally coherent activation sequences, or modal patterns that are the equivalent of autocatalytic sets for large-scale anthropogenic energy systems.

Modal Adaptation and Conversional Spectralization (MACS): The Relationship Between Technological Change and Growth

Ecology is the fundamental causal matrix behind the major historical changes in human society. The social and natural worlds are highly complex and nonlinear dynamical systems that are entangled together through intricate feedback loops. Recall that all economies are dissipative systems: they absorb vast amounts of energy from the natural world, use a portion of that energy to organize and sustain their internal structure, and dump most of the consumed energy back out to natural sinks in the global ecosphere. The ecosphere is capable of handling and assimilating a great deal of human waste and low-grade energy without getting severely destabilized. However, our current age of industrial capitalism is testing that proposition in every possible way. If the "useless" energy dumped by civilization into the natural world overwhelms the capacities of natural sinks and reservoirs, the result will be chaos throughout the global biosphere, much more so than what we've already seen in recent times. To help prevent that sobering possibility, we need to understand how the biophysical dynamics of technological change affect the scale dynamics of energy expansions and contractions. By stabilizing and modulating those scale dynamics effectively, we can indeed prevent most of the harmful impacts associated with the recent bionomic disruption while ensuring the prosperity of global civilization for millennia to come.

Let's begin with the ecodynamic theory of technological change and economic development, which works as follows. First, ecological primers cause changes in the underlying resource base or in the labor relations and modal patterns that have developed around existing resource bases. A resource base is any collection of natural primary energy supplies available for extraction, including things like fossil fuels, wind energy, solar energy, and water energy. A major example of ecological priming is the warming that occurred after the last Ice Age, which led to the proliferation of plants, animals, rivers, and lakes all over the world, thereby becoming the dominant factor in the transition from a nomadic existence to an agricultural one. Another example is the outbreak of a major epidemic, which can cause large death tolls and can change how the surviving individuals work and use technology. The Black Death is a major example of this kind of historical process, as the extreme labor scarcities it induced in Western Europe led to the proliferation of labor-saving technologies. Ecological primers and disturbances can change the underlying resource base through diversification or induced scarcities. When existing resources become scarce, societies typically try to compensate by trading with others, by switching to new resources, or by exploring and fighting to seize control of additional resources. When natural resources multiply and diversify, societies typically have more materials to collect, process, and manage. The new resource base typically has different energy qualities and characteristics compared to the old base, such as different energy and power densities or different spatial distributions. The change in energy quality then produces a change in modal flows and labor relations, mostly because people need to carry out modal adaptations to extract and process the new resources for predefined social needs. The rise of industrialization in England during the 1600s, 1700s, and early 1800s is a major historical example of this process, and one which I'll thoroughly cover in chapter 9. These broader ecological and social changes can induce many kinds of changes in technological development, but there's one in particular that is critical to the evolution of all economic systems: changes in the way that energy is converted.

I'll call this process *spectralization*, where spectralization is defined as the diversification and variation in the conversional methods of existing technologies in response to changing social and ecological conditions. A modern example of conversional spectralization has been the transition from gas-powered vehicles to battery-powered electric vehicles, since it involves a change in the way that cars convert energy as they move and perform other functions. Another example of spectralization was the emergence of wind turbines and solar panels for producing electricity, a notable change from the conversional methods used by coal power plants, which rely on burning coal to generate hot steam that drives a turbine. But not all changes to technological forms are conversional in nature. For about a decade, Ford's Model T cars were only painted black because that turned out to be the cheapest option. By the 1920s, increased competition from other automakers persuaded Ford to use a broader spectrum of colors. But the transition from black to colored cars is not conversional spectralization, because there were no changes in the underlying methods of converting energy. Another example of something that's *not* conversional spectralization are smartphone cases; they change all the time without any corresponding changes to the underlying technologies that power the operation of the smartphones. Likewise, logistical chains like roads and railways do not convert any energy themselves; they merely facilitate the conversion of energy by other technologies, such as cars and trains. Spectralization can happen along different economic scales and implementation vectors. On the scale side of things, we may speak of microlevel spectralization, which affects the conversional processes of specific technologies, or macrolevel spectralization, which may refer to conversional transitions for entire economies, like when trains and automobiles gradually replaced horses for personal transportation. At a conceptual level, it's also important to differentiate spectralization from the similar concept of technological *diffusion*. The latter concept is about the *spread and adoption of an already existing technological form* across society. By contrast, spectralization is about the *initial emergence and creation of new conversional technologies*. Another important difference is that although diffusion can apply to any technology, spectralization is something that applies specifically to new technological methods of converting energy. The two concepts may nevertheless overlap in significant ways on certain occasions, especially since the diffusion of preexisting technologies across society can affect the spectralization of new conversional methods.

The examples above are meant to reinforce the basic point that there are plenty of economic and technological changes that do not involve conversional spectralization. But even though not all technological innovations are spectralizations or even involve spectralizations, it's nonetheless true that spectralization is the core driver behind technological innovation because all economic systems are conversional networks, and thus changes to the conversional technologies that power those networks will have massive effects on the aggregate properties of the coronets themselves. The central idea behind spectralization is that conversional variety overcomes technical obstacles by channeling, modulating, and converting energy flows for the purpose of achieving different social objectives. People create tools, machines, and devices with new conversional pathways to overcome various energetic constraints and engineering challenges. Spectralized technologies are used to bypass chokepoints, bottlenecks, and reverse salients by synthesizing preexisting components and devices into novel combinations. The theoretical framework of modal adaptation and conversional spectralization (MACS) describes technological change as a chaotic and dynamic response to converging social and ecological conditions. As these conditions change, people respond by adapting and modifying their modal flows and labor relations. These adaptations can result either in changes to the applied methods on preexisting technologies, or to the gradual development of *entirely new technologies*, which arise as *contingent solutions to various problems that are discovered during the recursive and differential implementation of new modal patterns and flows*. Through this process of modal adaptation, people develop new biophysical methods for harvesting, extracting, and converting natural resources into finished products. Modal adaptation and conversional spectralization are the two fundamental elements of technological growth and innovation. In the MACS framework, modal adaptations and spectralizations form a synergistic feedback loop in which changes in adaptations can drive forward spectralizations and likewise changes in spectralizations can cause changes in adaptations.

The causal pathways by which spectralization overcomes energetic constraints can be highly diverse. For example, a device might be changed to convert energy more efficiently or a vehicle might be changed to transport more goods using the same volume of space, allowing people to save on production costs. But spectralization might also simply produce higher levels of exergy and useful energy *in the absence of any corresponding efficiency gains*. Likewise, spectralized technologies might produce *both* efficiency and exergy gains over time. Spectralization can thus lead to the expansion of an energy system through efficiency-driven gains, exergy-driven gains, both together, or through exergetic expansions that are enough to compensate for any corresponding declines in efficiency. Spectralization can boost efficiency and productivity gains because it gives people more control over the flow of energy. Economic incentives and social-institutional structures then provide selection pressures that result in the proliferation of certain conversional pathways over others, leading to standardized technologies and devices that come to dominate a conversional network, and therefore the economic system more broadly. The end result is that high rates of spectralization generally expand the scale of an energy system. However, forms of conversional spectralization that are carefully guided and controlled by the right mix of social factors do *not* necessarily need to produce larger energy systems. There's no inherent teleology or end-state in what spectralization does for society; it can produce different outcomes depending on how we guide it going forward. The spectralization of conversional methods and devices can lead to many kinds of changes across society, including changes related to infrastructure and built environments, such as the emergence of railroads after the invention of high-pressure steam engines or the construction of paved roads and highways after the invention of cars. Infrastructure often develops in response to the spectralization of conversional technologies, because society adapts logistical and transportation systems to facilitate the rise of a new conversional network.

Whether it's the rise of watermills and windmills or the development of steam engines and personal computers, the spectralization of technology has generally, though not always, produced more energy-intensive societies. To understand why spectralizations inflate the scale dynamics of energy coronets, it's useful to know that conversional methods have a high degree of entanglement with upstream and downstream economic activities. For example, a conversional process from any device requires energy inputs, since it involves the conversion of energy from one initial form to another. A different, spectralized conversional process may therefore require different inputs. Those new inputs have to be extracted, processed, or distributed using different methods, and executing those methods requires more energy. This extra throughput will raise the AFR of the economy unless there's corresponding downscaling effects in other sectors. What's more, the outputs of the spectralized process may also change, and they would then need to be distributed and consumed via different modal adaptations across the nodes of a given coronet. Of course, spectralization begins at the level of individual devices and components. But these microlevel spectralizations can profoundly affect the scale dynamics of conversional networks if they manage to broadly diffuse across the economy. Under modern capitalism, this process of diffusion and technological management unfolds because of various social selection mechanisms guided by the dynamics of class conflict, social power, and geostrategic rivalries. This process is described in much greater detail in chapter 8.

To better understand the effects of spectralization on energy scale dynamics, let's consider a concrete example: the macrolevel transition to renewable energy sources like wind, solar, and hydro. As of 2021, renewables provided an impressive 28 percent of the world's total electricity generation, up from a share of 20 percent just eleven years earlier.[10] The rate of growth in the last few decades has been astonishing. This macrolevel transition in the global economy is being driven by numerous microlevel spectralizations involving technologies like electric batteries, wind turbines, and solar panels. It's also changing the input and extraction dynamics on which the global economy depends, as the raw inputs to the new coronets are shifting from fuel-based sources to material-based sources. Renewable technologies rely heavily on critical minerals like cobalt, lithium, nickel, copper, and rare earth elements, including yttrium, scandium, and the lanthanides.[11] According to the International Energy Agency, a typical electric vehicle requires six times the mineral inputs of a conventional gas-powered vehicle.[12] The typical onshore wind plant needs nine times more minerals than a conventional gas-fired power plant.[13] Since 2010, every new unit of power generation has required on average 50 percent more mineral resources.[14] But despite the impressive scale of this transition, the world is still using more energy, emitting more greenhouse gases, and setting new records in the atmospheric concentration of carbon dioxide and other greenhouse gases. That's in large part because the transition to renewables is still heavily dependent on fossil fuels, as raw minerals are mined and transported using vehicles powered by fossil fuels. Likewise, wind turbines and solar panels are still largely built in factories operating on electricity generated by coal-fired and gas-fired power plants. Furthermore, distributing all the new electricity produced by wind and solar farms will require the installation of additional grid capacity. This distributional barrier has been a major problem in the United States, as thousands of renewable energy projects have been delayed because of bureaucratic mismanagement and an insufficient electric grid.[15] Fewer than 20 percent of all proposed wind and solar projects in the United States actually come to fruition.[16]

Microlevel spectralizations affect the individual devices that interact within a broader coronet, such as prime movers. In a general sense, a *prime mover* is any initial source of motive power, such as a windmill or an internal combustion engine, that receives a flow of energy and then uses that energy to drive machinery or perform some other tasks. In engineering specifically, prime movers are engines that convert fuel to useful energy. For our purposes, I'll define prime movers as the dominant energy converters of a given economy—in other words, they're the most energy-intensive devices operating in the economy. Prime movers have changed radically over time, as documented by Vaclav Smil in *Energy and Civilization*. For much of human history, human muscles provided the peak power capacities available, at roughly 100 watts (W). Draft animals like oxen and horses reached about 400 W in the third millennium BCE. By the year 1000 CE, horizontal waterwheels had become the leading prime movers, generating about 5,000 W of power. In 1800, the biggest steam engines had blown through 100,000 W and retained their dominance throughout much of the nineteenth century. In more recent times, steam turbines have become the dominant prime movers, with the biggest units producing more than 1 GW, or one billion watts. In contrast to prime movers, *contingents* are relatively low-energy devices that depend on prime movers in order to successfully convert energy and operate normally. A personal computer is a great example of a contingent device. It operates in part because of massive steam turbines producing vast amounts of electricity, which is then distributed through various modal flows to the different residential, commercial, and industrial districts of an economy. Personal cars and vehicles are another common example of contingent technologies; they can only operate because of highly processed fuels or electric charge carriers that have undergone multiple stages of conversion in their modal flows across the economy. Industrial machines can also be considered contingents, and in that sense contingents act on raw materials to create finished manufacturing products that are sold in regional and global markets.

The spectralization of prime movers can have important downstream effects on economic activity, as prime movers effectively dictate the exergy capacity of the whole system. However, the spectralization of contingents can also profoundly affect the spectralization of prime movers. That's because technology exists in the context of financial and commodity markets that are controlled and engineered by certain groups to yield certain results. The modal flows and dynamical cycles of modern capitalism are therefore organized for the purpose of producing and delivering large volumes of contingent devices and products in a timely and efficient manner.

In the past two centuries, we've lived under an energy-intensive technological regime of catalytic spectralization. The spectralization of technology under capitalism is catalytic in two primary ways. First, increasing spectralization has the effect of accelerating and intensifying modal flows across the coronets of a given economy.

In other words, conversional processes under capitalism are modified to enlarge and speed up cycles of production and distribution. That's because spectralization generally leads to a greater demand for energy services to build, operate, and maintain the spectralized technologies. In the meantime, older technologies or energy systems don't go away; they're simply shifted and transferred for the purpose of achieving other tasks and objectives. Modern civilizations no longer need horses for transportation, where ground vehicles and airplanes have taken over, but horses are nevertheless still used for competitive racing and other commercial events. Second, capitalist economies are generally catalytic in the sense that technological innovations in one economic sector often catalyze and speed up changes and innovations in other sectors, thus expanding the energy scale of the overall system. For a common example, if one particular company or economic sector spectralizes a particular technology that happens to reduce production costs, other companies and sectors will inevitably follow in order to stay competitive. The result is not just that spectralized technologies diffuse across the economy and therefore consume more energy, but that an increase in the rate of spectralization itself becomes a major feature of economic development under capitalism. Even in the presence of economy-wide efficiency gains, higher rates of spectralization under capitalism will increase the aggregate energy scale of our civilization, thus imposing greater pressure and more disruptions on the natural cycles of the biosphere. These intense disruptions will then reverberate on human civilization itself, causing it to buckle and bend from all the collective ecological pressure that's rapidly building up.

Efficiency and Technological Innovation

Many people today believe that we can carefully manage and mitigate the ecological crisis we face through clever technological innovations. One of the more popular arguments inside the circles of technocapitalism is the idea that civilization can just keep humming along through gains in energy efficiency. The idea that we can do more with less is seductive, but there are several reasons why this strategy will fail over the long run, if we choose to keep pursuing it. The most fundamental reason is that nature imposes absolute physical limits on efficiency that no amount of technological progress can overcome. The recent breakdown in Moore's Law because of quantum effects is a notable example.[17] Another one is the efficiency barrier that the Carnot cycle represents for all practical heat engines.[18] Second, even if aggregate efficiencies improve through amazing technological innovations, they're likely to do so too slowly given the timing constraints on implementing the new and ambitious energy policies required to get us out of our current mess. And third, aggregate efficiency gains for entire economies are almost always associated with *higher* levels of energy use and consumption, not less. This last claim appears to be quite paradoxical, so let's explore it first.

We sometimes drive longer distances when fuel efficiency improves. We often power more appliances when electricity becomes cheaper. Those who are proud of energy savings at home, through recycling and other similar activities, are more than happy to take vacations by jumping on an airplane and flying halfway around the world. People often take energy savings in one area and exchange them for expenses in another. What we end up doing with energy efficiency gains can sometimes be just as important as the gains themselves. In ecological studies, this phenomenon is generally known as the *Jevons Paradox*, which reveals that the intended effects of efficiency improvements do not always materialize.[19] First formulated in the nineteenth century by the British economist William Stanley Jevons, the paradox states that increases in energy efficiency are generally used to expand accumulation and production, leading to greater consumption of the very resources that the efficiency improvements were supposed to conserve. The argument behind the paradox is that boosting efficiency leads to cheaper goods and services, which encourages more demand and more spending, leading to the consumption of more energy.[20] Jevons described this effect in the context of coal power and steam engines. He observed that efficiency improvements in steam engines had encouraged more consumption of coal in Britain, implying that boosting energy efficiency did not actually lead to energy savings.

In economics, variations of this paradox are known as the *rebound effect*. For example, if a 5 percent rise in fuel efficiency leads to a 2 percent decline in fuel consumption, then the rebound would be 60 percent. The Jevons Paradox, also called "backfire" among economists, would occur if a rise in fuel efficiency actually produces an increase in fuel consumption. Most neoclassical economists accept that some rebound effects are real and significant, but they largely reject the notion of backfire. Neoclassical studies have concluded that microeconomic rebound effects hover around 20 to 40 percent whereas macroeconomic rebound effects are much larger, at around 50 to 60 percent, though in some cases they could reach even higher.[21] It should be emphasized that the results of these studies are not very useful, for several reasons. First, they're highly sensitive to the time horizon under consideration. Government policies or autonomous efficiency improvements might lead to net energy savings for ten years or so, then completely backfire fifty years down the road. Simply put, the studies make assumptions about demand patterns in the economy that are unlikely to hold for long periods of time. Second, these studies are generally terrible at discerning boundary effects. Efficiency policies or improvements might lead to net energy savings for one particular company, or in one particular country, but those same improvements can then spill over and produce higher energy consumption among other companies and countries. As energy scientist Carey King puts it: "The smaller the system boundary used for analyzing the backfire effect, the less relevant the paradox appears. To fully conceptualize the Jevons Paradox, your boundary must include the entire world economy, or every single device that consumes energy."[22] In a comprehensive review of the literature on the subject, the moderate UK Energy Research Centre claimed, rather prematurely, that the most extreme versions of the rebound effect probably no longer apply to developed economies. However, the research group also argued that large rebound effects across our economies can still occur. They reached the following conclusion: "It would be wrong to assume that ... rebound effects are so small that they can be disregarded. Under some circumstances, such as energy efficient technologies that significantly improve the productivity of energy intensive industries, economy-wide rebound effects may exceed 50 percent and could potentially increase energy conversion in the long term."[23] Even if one conceded that backfire is impossible, the mere fact that significant economy-wide rebound effects are indeed possible should give all of us pause about the utility of efficiency strategies in combating the ecological crisis and climate change.

For all these nuances, much of this argument is actually a red herring. The fundamental problem is that the entire debate obscures a more important unknown: the problem of whether efficiency improvements, even if they don't backfire, can come fast enough to alleviate the worst consequences of the ecological crisis, which are still ahead of us. In other words, the central problem in this discussion is about time. Can we make huge efficiency gains in the short period of time required to take serious action and forestall some of the more devastating consequences of the bionomic disruption? Do we really want to stake the future of global civilization on the fringe promise that massive efficiency gains on a global scale can be realized within a couple of decades, contrary to all historical evidence? Given the economic and political incentives prevalent in our current age of capitalism, it's unlikely that our love affair with efficiency can actually provide concrete solutions to the profound challenges that we face. That's because the aggregate efficiency of an economic system is an inertial quantity that changes at a glacial pace, regardless of whether it's increasing or decreasing.

We see this exact process playing out right now with greenhouse gas emissions, although our ecological crises extend far beyond this problem. Political and business leaders have been hoping for years that technological progress will somehow deliver both higher rates of economic growth and a sharp reduction in greenhouse gas emissions. Things have not gone according to plan. The United Nations has long warned that an "unacceptable" gap exists between the pledges from national governments and the emission reductions needed to prevent some of the worst consequences associated with climate change.[24] In 2022, carbon dioxide emissions reached a record high globally, completely defying even the relatively modest goals of the Paris Agreement.[25] Unsurprisingly, atmospheric levels of carbon dioxide, methane, and nitrous oxide—collectively the most dangerous greenhouse gases—also reached new highs in 2022, setting the stage for even more global warming down the road. The challenges with boosting efficiency are easier to understand when we view capitalism on a global scale: although many developed nations have made modest yet measurable improvements with their aggregate efficiencies, these gains have been completely washed away by developing economies that are still in the process of industrialization.[26] Evidently, substantial changes in the aggregate efficiency of the global economy are unlikely to materialize in short periods of time under the capitalist regime. Technological growth under capitalism will deliver some additional progress on efficiency, but certainly not enough to prevent the worst consequences of our ecological crises.

One of the best ways to understand the inertia of aggregate efficiencies is to compare energy efficiencies under capitalism with those from nomadic days, more than 10,000 years ago. Recall that human muscles performed the largest share of the work in nomadic societies, and the efficiency of our muscles is roughly 20 percent, perhaps much more under special circumstances.[27] For comparison, most gasoline-powered combustion engines have a thermodynamic efficiency of roughly 15 percent, coal-fired power plants come in at a global average of about 30 percent, and the vast majority of commercial photovoltaics are somewhere around 15 to 20 percent.[28] Variations exist for all of these numbers, depending on a wide array of physical conditions. Nevertheless, when it comes to efficiency, we can safely conclude that the dominant prime movers and contingents of capitalism can hardly do much better than human muscles, even after three centuries of rapid technological progress. Large gains in efficiency are extremely difficult to achieve, in both physical and economic terms, because they require enormous investments and resources. Although impressive efficiency gains happen occasionally in the history of capitalism, they have always been subordinated to efforts at expanding useful energy output and the scale of production more broadly.

From time to time, amazing innovations come along that set new standards for future technologies, but an amazing innovation does not represent the entire economy. The Watt steam engine was a major improvement over previous models, but its thermal efficiency was only 5 percent at best and its diffusion across the English economy was rather slow.[29] For another example, Tesla motors themselves have a phenomenal operating efficiency, but the electricity needed to run the cars often comes from much more inefficient sources, such as coal power plants. In 2013, a person driving a Tesla in Ohio or West Virginia would have produced roughly the same carbon emissions as someone driving a Honda Accord, given the dirty energy mix in those states.[30] The aggregate efficiency of modern economies remains relatively low because elite capitalists are interested in increasing their profits and production levels, not in making the enormous investments required to generate significant improvements in efficiency. As long as the current economic order remains in place, capitalism will continue to thrive on energy-scale expansion and intensive dissipation, with efficiency improvements merely secondary and incidental to the wider project of plundering the natural world and profiting from that plunder. Since efficiency gains alone are unlikely to improve the outlook of global civilization in the face of the bionomic disruption, we need to think more comprehensively about the future direction of technological change and what technology should be doing for civilization.

The Future Direction of Technological Change

To wonder about the future of technology is to wonder about the future of society as a whole. There's no shortage of opinions and narratives about the future direction of technological innovation.[31] There are multiple camps in these debates, and the debates themselves unfold along multiple vectors. Perhaps the most prominent vector is directed at the underlying resource base that human civilization should adopt in the future. For debates along this vector, there's one side that has concluded that fossil fuels are absolutely necessary for the future success of human civilization, and therefore the proper way to combat their harmful effects is by directly intervening in the ecosphere, either through solar radiation management or through widespread greenhouse gas removals. These ideas are collectively known as *geoengineering*, and I'll say more about the concept shortly. Another camp argues that we need to implement a rapid and radical transition away from fossil fuels and toward renewable energy sources, such as hydro, wind, and solar. Others maintain that we can follow a kind of hybrid approach where we develop both fossil fuel and renewable technologies, the "everything goes" strategy. Another vector of debate is directed at the role that technological innovation should play as a response to our ecological crisis. The most popular camp in this debate believes that humans are eternally clever and resourceful; combine that ingenuity with the social mechanism of capitalist markets and civilization can then always extricate itself from any crisis. What most of these debates have in common is that they're happening within the ideological framework of capitalist economies. In other words, they assume that capitalism is the underlying economic system we should aim to preserve, and what really matters is the technological parameters under which capitalism develops in the future. For the most part, they're very narrow debates about technological tinkering, leaving aside the more important issue of what political and economic regimes are going to manage, supervise, and direct these various technological paths.

The reason why elites frame the potential solutions to our common global problems as a simple matter of technological tinkering is simply because that's what would allow them to preserve their wealth and power, to preserve the status quo from which they benefit. Capitalists are the apex predators of the ruling classes, and capitalists anywhere and everywhere are masters of deception and distraction. Focusing the debate on the Promethean potential of technology has become a specialized ritual for sidestepping more difficult conversations about the structural distribution of power in modern society. By now it's hard to keep track of all the technological breakthroughs that are supposed to save us. Maybe it's wind and solar. Maybe it's carbon capture and storage. Maybe it's solar radiation management. Maybe it's ocean fertilization. Maybe it's next-generation nuclear power plants. Maybe it's high-temperature superconductors. Maybe it's hydrogen fuel cells. Maybe it's nuclear fusion. Maybe it's electric vehicles. Maybe it's solar panels in space beaming energy down to Earth. Maybe it's all of them. Maybe it's something else. Maybe it's Elon Musk. Maybe it's Santa Claus or Harry Potter. Fictional characters have just about the same odds of getting the job done as all these other things. It's impossible to cover the intricate nuances of all these proposals in a single chapter, so here I'll focus on two big things: renewable technologies and geoengineering.

Among those who have correctly decided that we eventually need to ditch fossil fuels, at least for the majority of our energy production and economic activities, there are still further debates around what should replace them. The two broad camps in this debate are those who support renewables like wind and solar and those who support nuclear power as the salvation strategy for our civilization. There are many critics who contend that wind and solar are too unreliable for the energy demands of a modern electric grid.[32] They're supposedly "intermittent" sources of energy, and the implication is that other sources, like nuclear power plants and coal-fired plants, are more stable. The reality is very different. All sources of energy are intermittent at different timescales.

Let's consider France, a nation that produces roughly 70 percent of its electricity from nuclear power, a higher share than in any other country.³³ In 2019, every French nuclear plant was shut down on average for almost 100 days because of planned repairs or other emergency issues; that figure rose to 115 days in 2020 when French nuclear plants generated "less than 65 percent of the electricity they theoretically could have produced."³⁴ In 2022, the power output of the French nuclear industry fell below 50 percent capacity, reaching a thirty-year low.³⁵ Chronic underinvestment from Électricité de France (EDF), the state-backed company running the nuclear power system, is a major contributing factor to the recent failures, but the sheer complexity of nuclear power doesn't help. Many French nuclear plants are suffering from corrosion and faulty welding seals. Furthermore, as I mentioned in a previous chapter, many nuclear power plants around the world are heavily reliant on cool rivers to keep their reactors in check, and global warming is going to make those rivers warmer and warmer as time goes on, thus sabotaging electricity generation from much of the nuclear industry. This is not a hypothetical problem; it's a major issue that many countries, including France, have already experienced.

Nuclear power advocates often tout the advanced features of nuclear technology, but that technological complexity is actually a great reason *not* to rely on nuclear power as a substitute for fossil fuels. Precisely because nuclear power plants are so complex to design and build, they are plagued by ridiculous cost overruns, construction delays, and challenging operational and maintenance issues that last throughout their lifetimes. For example, EDF builds nuclear plants in other European countries. It built one in Finland that started operating in 2022 but was supposed to be ready *thirteen years earlier*, in 2009. The nuclear industry in the United States has suffered from the same problems. The planned expansion of the VC Summer nuclear power plant in South Carolina went disastrously after its main contractor went bankrupt in 2017.³⁶ The expansion has since been scrapped, with nothing to show after billions of dollars wasted, and the entire affair eventually devolved into accusations of fraud, lawsuits, recriminations, and various legal settlements. The expansion of the Vogtle plant in Georgia has experienced some of the same issues. Everywhere you look around the world, the nuclear industry is in shambles and can't get its act together.

In February 2021, when a massive winter storm knocked out power all over the U.S. state of Texas, wind and solar energy sources generally outperformed their fossil fuel counterparts.[37] Indeed, the scale of the disaster was amplified largely because of frozen natural gas pipelines and a lack of weatherization among power plants using fossil fuels. Looking at the System Average Interruption Duration Index (SAIDI), an indicator that measures grid reliability, also confirms this point. In 2020, Germany had a SAIDI of just 0.25 hours, one of the lowest in Europe.[38] By contrast, the United States, where nuclear power provides 20 percent of electricity, had a SAIDI of 1.28 hours in 2020, roughly five times the outage rate of Germany. In contrast to the typical drivel that surfaces on this issue, renewable energy sources are indeed a highly reliable source of energy and can be successfully used to manage intermittency issues when fossil fuel and nuclear power plants become unavailable. Just as important, multiple studies have shown that the broad adoption of nuclear power would lead to much higher greenhouse gas emissions than the broad adoption of wind and solar. It's certainly true that the nuclear industry would have a smaller carbon footprint than the current fossil fuel industry does, but the uncertainties surrounding future emissions from the expansion of nuclear power are huge.[39] It's almost certainly the case that the diffusion of nuclear power would produce much larger GHG emissions than the corresponding diffusion of wind and solar.[40] Given all these problems, it's fairly obvious that nuclear power should not be our main strategy going forward, although in highly select cases nations should still continue to build some nuclear power plants to supplement the core strategy focused on the expansion of renewable energy sources.

The most fruitless debates, however, are the ones involving supporters of the current fossil fuel regime. For this crowd, there's no sense of urgency and no energy transition required to deal with our ecological problems. Instead, they would deal with these problems by implementing geoengineering solutions designed to cool down the planet and suck out greenhouse gases from the atmosphere while at the same time continuing to intensively burn fossil fuels. Geoengineering is large-scale anthropogenic intervention in the natural systems and energy flows of the global ecosphere for the purpose of managing the harmful effects of our economic activities. Overall, there are two broad geoengineering proposals on the table: solar radiation management (SRM) and carbon dioxide removal (CDR). The basic goal of SRM strategies is to somehow change the chemical composition of the atmosphere or the planet's surface so that more sunlight is reflected back into space. A prominent proposal is to inject stratospheric aerosols—small particles that reflect sunlight—into the upper atmosphere. There are several big red flags associated with SRM strategies. Perhaps the most serious is that they risk severely destabilizing the global hydrological cycle, which would lead to massive droughts around the world.[41] Another major problem is what's known as the termination shock, the point in time when we have to stop doing SRM. That's going to cause a huge and sudden shock to global temperatures, because once we pull back whatever we're doing to reflect sunlight back into space, much more sunlight will start hitting the surface again and temperatures will go back up almost right away. One might wonder why have to stop doing SRM at all. Why can't we just keep going with it forever? The answer is related to the first point above. Sooner or later, continuing to engage in SRM will virtually destroy the natural cycles of the biosphere, so we'll have no choice but to stop it or roll it back significantly, which will inevitably produce a termination shock that wallops global civilization.

In contrast to SRM, CDR strategies are focused on removing carbon dioxide from the atmosphere. A major example of a CDR proposal is afforestation, which would require a global effort to plant as many trees as possible, thus expanding the natural ground sinks for carbon dioxide. Countries such as China have had some limited success with this strategy, but it's doubtful it could work quickly enough for the entire world. Another idea is ocean fertilization, a strategy that calls for people to add critical nutrients in special locations around the world's oceans as a way of promoting the growth of microorganisms that would then suck out even more carbon dioxide from the atmosphere. This strategy has proven to be mostly unreliable, mainly because the vast majority of the carbon sucked up by the ocean would stay near the surface, therefore making it likely to end up back in the atmosphere.[42] Very little carbon would reach deep into the ocean where it could be naturally sequestered. And yet another example, perhaps the most prominent one, is carbon capture and storage (CCS), which refers to the process of building specialized facilities or installing devices in power plants that are designed to remove greenhouse gases from the atmosphere. CDR proposals now form a major component of all IPCC models about future climate scenarios. There's a popular assumption among politicians and capitalists around the world that we can just keep burning fossil fuels with reckless abandon, because we can always just pull out that carbon from the atmosphere later or prevent it from getting there in the first place.

One of the more popular CCS strategies has been bioenergy carbon capture and storage (BECCS). It works by burning cultivated crops to generate electricity and then capturing the resulting carbon dioxide and sequestering it underground, so that it never gets into the atmosphere. As of 2023, BECCS facilities were capturing roughly two megatons of carbon dioxide per year, a trivial amount compared to the roughly 50 gigatons of greenhouse gases that are emitted globally every year. According to the International Energy Agency, carbon removal through BECCS facilities will reach roughly 40 megatons a year by 2030, which is far less than the 250 megatons a year envisaged in the IEA's own Net Zero by 2050 scenario.[43] Beyond the fact that BECCS implementation is ridiculously behind schedule, it's worth questioning whether that implementation is a good idea in the first place. Implementing BECCS on a global scale could be disastrous for global civilization as it would require vast amounts of new lands, roughly the size of Australia, to be set aside for crop cultivation and electricity generation. It would also be extremely costly. The climatologist James Hansen has estimated the financial costs of setting up BECCS in this century to be hundreds of trillions of dollars.[44] And because BECCS will likely be established in farmlands with crop monocultures dominated by big agribusiness, there will almost certainly be massive amounts of new greenhouse gas emissions associated with harmful land use changes. The traditional counterpart to BECCS is simply CCS, which is designed to capture and store carbon dioxide at power plants. It suffers from some of the same basic problems. Regarding the implementation of CCS, energy scholar Vaclav Smil has concluded that "in order to sequester just a fifth of [2010] CO_2 emissions we would have to create an entirely new worldwide absorption-gathering-compression-transportation-storage industry whose annual throughput would have to be about 70 percent larger than the annual volume now handled by the global crude oil industry, whose immense infrastructure of wells, pipelines, compressor stations, and storage took generations to build."[45] Overall, it's safe to conclude that CCS strategies are little more than carefully orchestrated propaganda from the fossil fuel industry designed to prolong their profits and the current economic order from which they benefit.

Given all these challenges, a renewable strategy based on the "Holy Trinity"—wind, solar, and hydro—is the most plausible and realistic path forward in the future. Whether humanity takes it is another matter. That these three should be the core components of our future energy strategy does *not* mean that other energy sources cannot be used. And if we're going to implement this strategy, then the way we decide to deploy and develop it needs careful consideration. Because it takes energy to produce more energy, scaling up new coronets based on renewable energy sources could also end up becoming vastly energy-intensive, unless the transition to renewables is properly managed and constrained through a political and economic process rooted in the realities of the planetary biosphere. For example, transitioning to renewables will require using additional land, constructing large solar and wind farms through complex machinery, the mining and production of new minerals and elements, and the expansion of transportation networks through global trade. These changes will need to be balanced by downscaling effects in other economic sectors, because in the absence of any strategic constraints from state power, this vicious energy-technology spiral would simply aggravate our current civilizational problems even more. A possible countervailing constraint is to implement a vast reduction in global beef production, which requires enormous tracts of land for the cultivation of crops that are used to feed cattle. This is just one possible change; the larger point is that there are many complex dimensions to land use dynamics that humanity would have to think through as it carries out an energy transition away from the fossil fuel addictions of modern capitalism. A radical and rapid transition toward renewable energy can and must take place, but only if it's accompanied by massive cutbacks in aggregate energy use via the decommissioning of more coal-fired and gas-fired plants and the political imposition of new limits on the transportation sector, along with other forms of industrial policy, to manage and ensure the stability of market prices and other financial indicators.

For another example, hydroelectric dams operating in warm climates and low altitudes actually emit large quantities of greenhouse gases, especially methane, in large part because of decaying organic matter confined in the water reservoirs. Some hydropower facilities produce more greenhouse gas emissions than coal-fired power plants. If we're going to expand renewable energy capacity by building additional hydropower facilities, then we should focus on building them in cooler climates and higher altitudes, where water reservoirs are likely to hold fewer organic materials capable of decomposing and of releasing harmful gases like methane. That's why Norway's hydropower plants, which produce roughly 90 percent of the country's electricity, have extremely low life-cycle GHG emissions comparable to typical solar and wind farms.[46] It makes sense to build hydropower capacity in places like Norway, Russia, and Canada. But from the perspective of a stable biosphere, it's much more difficult to do it in places like Brazil, which has long toyed with the idea of constructing dams along the Amazon. That's not to say it's impossible. But if it's going to be done in an ecologically beneficial way, it needs to be much more carefully planned than it has been so far. It's also important to remember as we decommission fossil fuel plants in the future that they do have some advantages over wind and solar, such as flexibility and demand management. If there's too much electricity in the grid, we can simply turn off a coal plant and it won't produce anything. Even with significant improvements in energy storage technologies, it's harder to generate the same level of flexibility with wind turbines and solar panels. Thankfully, hydropower offers precisely that level of maximal flexibility because any hydro plant can be toggled on and off as necessary. What's more, pumped hydro storage is one of the best-established methods of energy storage and is rapidly spreading around the world.[47] It works by pumping water from one reservoir to another reservoir located at a higher elevation, then storing the water there until it's needed later. At that time, the water is released back to the lower reservoir, driving a turbine that produces electricity along the way. The best part is that pumped storage doesn't even need natural rivers or lakes to work. Any site that can store water will do.

The other major consideration is that even though we need to rapidly downscale the fossil fuel industry, it's not a good idea to eliminate it right away. One reason is because the pollution generated by burning fossil fuels results in layers of particles in the atmosphere that prevent some solar radiation from reaching the surface of the planet. That process has a cooling effect on global temperatures.[48] If we turned off all refineries and fossil fuel power plants tomorrow, global temperatures would rapidly accelerate upward for a short period of time, leading to chaos in the global economic system. Another major constraint on the immediate elimination of fossil fuels is transportation. Although electric batteries for ground-based transportation have made enormous strides by now, we don't have commercially viable equivalents for air transportation or merchant shipping. The Chinese company CATL, the world's dominant EV battery manufacturer, did announce in early 2023 that it would begin mass producing a next-generation battery with an energy density that could reach up to 500 watt-hours per kilogram, roughly double the energy density of the best batteries on the market.[49] CATL also claimed that this battery could be used in passenger aircraft. But even if that's the case, it could take years for the technology to be considered reliable enough for widespread adoption in the airline industry. And that's why targeted spectralization is so important; we need different social management strategies for different industries. Electricity generation can undergo a radically quick transition toward renewables; there's no technological barrier there at all. It's purely social and political barriers that are the problem. But we have to be careful with transportation, because there the technological barriers are indeed more serious.

In contrast to the catalytic spectralization of late-stage capitalism, we need strategic, balanced, and targeted spectralization that harnesses innovation in critical industries but carefully manages technological change in others. Under the capitalist regime, it's easy for network and market effects to rapidly escalate, driving up energy use to extremely high levels and therefore dissipating more energy to our external environments. If we want to target technological innovation in a few industries, it has to be balanced by countervailing constraints in other industries. Achieving such targeted spectralization cannot happen in the context of the existing power relations in late-stage capitalism. We need new political and economic institutions that prioritize the long-run stability of our biosphere as well as the economic concerns of workers worldwide. And to get there, we'll need to better understand the complex relationship between society, energy, and technology—that is, to understand how the actions of the social institutions that govern our lives affect the energetic and technological dynamics that are increasingly governing the global ecosphere.

CHAPTER 8

Energy and Technology in the Social Sphere

People want to extract useful energy from their technological systems and the natural world, but what's useful can only be determined by emergent social conditions. Sunny days at the beach are good for local hotel owners because it means that more people are likely to stay at their hotels, boosting revenues and profits. But the hotel owners do not control the Sun or the weather; they are at their mercy, just like the rest of us. There are many natural energy flows that could be useful to human beings in various social and economic situations. However, our history as a species shows that we are not content to sit around and wait for nature to do things for us. We have learned not just to admire what nature gives us, but to change and manipulate the gifts of nature for the purpose of doing specific tasks and solving specific problems. We took wood and stones and turned them into sharp spears so we could hunt wild animals. We dug canals next to big rivers and irrigated our farms to produce more food. We used metals and fossil fuels to build heat-powered engines so we could lift more things and travel faster over long distances.

One way or another, technological innovation has profound social and economic dimensions. People don't improve technology just to improve technology; they improve technology to achieve some other aim, whether it's because they think better technology might help them win a war, seize control over a particular market, or some other reason. A global tech company like Apple does not invest billions of dollars in research and development because of some altruistic goal to improve humanity; it pours those billions into improving its smartphones because it wants to sell the resulting commodities in global exchange markets, boosting profits as much as it can. In this sense, corporate power, class struggles, and political conflicts are also involved in guiding the direction of technological development. The rich and powerful use technology not simply so they can produce more stuff to sell while underpaying their workers, but also to establish and reinforce mechanisms of control over society, thus allowing them to extend their rule. Technology should never be analyzed separately from the social and class dimensions of our economic systems; all technological innovations are embedded in the wider social web in which we live and work.

The fact that social and institutional forces constrain the process of technological change does not mean that ingenuity and creativity no longer play an important role. Technological growth does not happen simply because of greater funding for science and other research departments. It also happens in the normal, day-to-day activities and practices involving technological products. A blacksmith who's been making swords for years may find a new technique for making them faster, or stronger. A plumber who's been fixing toilets and faucets his entire life may notice something new in the process of fixing a single item, then launch a new invention on the basis of that experience. A repairman fixing a broken part on a bicycle might notice that the part would break less often if it was installed in a different way or in a different location. He could then go and make his own bicycle with the new configuration, then he could patent it to make sure he's the only one who can successfully commodify his idea. The company that he establishes based on this patent will make many other types of bicycles and sell them for profit, or at least try to. At that point, technological innovation will be increasingly subject to the financial demands and political constraints of a capitalist economy: the need to score higher profits than your rivals, the need to keep shareholders happy, the need to sabotage workers by limiting wages and benefits, and the need to enter new markets.

The Class-Power Dynamics of Technological Change

Social hierarchies act as a powerful filter for the transmission and distribution of technological innovations, and understanding these social dimensions is critical for building a society where technological development is steered toward the benefit of the broader public and the planetary biosphere. This lesson is especially true when it comes to how people use and extract energy through technological systems. In this book, when I talk about the release or extraction of *useful* energy, it is only in the context of technological systems built and designed by humans. Although natural energy flows are useful in many contexts on their own, the main focus here is on the way that we use various biophysical systems, from animals to jet engines, in order to change, convert, and filter those natural energy flows for some specific social purpose. In this sense, technology encompasses physical products, but it also includes *social ideas*, or the thoughts and beliefs that are necessary for sustaining and altering the course of technological development. As I explained in the prior chapter, technology has no teleological destiny apart from the dynamic social and ecological conditions in which it's embedded. It has no fixed and predetermined goal, such as always aiming for greater efficiency. Technological change is immanent and contingent, meaning that technologies develop in response to various social and natural conditions. This claim does not deny that technological improvement can sometimes occur simply because of accidents and fortunate circumstances. That happens often. But the wider contours of energy production and technological development are established by those who have the social and institutional power to do so, and that's the social process that we'll turn to next.

Marx and Engels start *The Communist Manifesto* with the following thunderous lines:

> The history of all hitherto existing society is the history of class struggles. Freeman and slave, patrician and plebeian, lord and serf, guild-master and journeyman, in a word, oppressor and oppressed, stood in constant opposition to one another, carried on an uninterrupted, now hidden, now open fight, a fight that each time ended, either in a revolutionary reconstitution of society at large, or in the common ruin of the contending classes.[1]

Marx and Engels are describing human history as a dialectical and dynamic social process in which competing economic classes and forces are the main drivers of historical change, in contrast to Hegel's idealistic notion of Spirit and reason as the central actors of history. The concentration and deployment of massed labor power is the fundamental social source of production, and in every age of civilization there arises a class system in which a small fraction of society takes strategic control over the labor power of the vast majority of the population, effectively telling the latter what to do and how to think. But classes are never frozen in time and place; history is a meandering, flowing river. Whether because of conflict, cooperation, or a million other factors, society is always engaged in a process of dynamic transformation. This historical process of transformation involves wars, revolutions, cultural creativity, technological innovations, medical marvels, ecological disasters, and so many other social and natural processes. As the class dynamics and labor relations of society shift in response to changing ecological and historical conditions, they carry along with them virtually all other aspects of social existence, including the nominal domain of money and finance.

When most people think about finance and economics, money is usually the first thing on their minds. Energy might be the foundational cornerstone of economics, but our social lives are dominated by money almost everywhere we go. We don't pay for college or for groceries with units of energy; we pay for them with units of currency officially approved by governments and often issued by established institutions like major banks. The economist Warren Mosler is fond of telling a story about his children.[2] One day he informed them that he would give them his business cards if they would help clean up around the house. Nothing really happened; the children ignored his bizarre request because Mosler's business cards were useless to them. But then he changed the rules of the game, since he had the power to do so: he told them that if they did not perform enough chores and collect enough business cards by the end of the month, he would ground them all and they wouldn't be able to see their friends. Mosler had effectively created his own currency; his children suddenly needed to collect his business cards by doing specific chores, and each chore got rewarded with a certain number of cards, because otherwise they couldn't leave the house! This example highlights how money functions as a social relation, but it also highlights the important role of power and hierarchy in making it all work. Mosler could only pull off his trick because he was the parent and he established the rules.

Money is a social creation, a complex set of social conventions used by people to organize their relations and hierarchies. To say that money is a "social" phenomenon means that it's a complex set of social and cultural conventions, as opposed to being a law of nature that human beings do not control. As a social creation, money is something over which people can plan, negotiate, and bargain. The dynamics of money can be changed and altered through social interactions. Debts can be created and forgiven. Salaries can be raised or reduced. Profits can vanish and reappear. And just like energy flows through highly structured coronets, the flow of money through our economies is strictly controlled by dominant classes and powerful nations in the wider global system. The nominal domain of prices, wages, debts, and profits largely functions as a symbolic quantifier of class struggles and social power dynamics. These relations are entangled in a complex web of causation involving the material sphere of production and other biophysical factors. That's because the large-scale domination and exploitation of labor by the ruling classes requires vast amounts of energy, as the entire technosocial structure of human civilization needs to be dynamically organized and transformed to preserve the current order, to keep it in a state of equilibrium against disequilibrating tendencies in society and more broadly in nature. And since capitalism has taken the drive for financial accumulation to unprecedented levels, it needs to impose ever-increasing systems of control, surveillance, and automation, all of which require huge energy expenditures that will impose brutal costs on the ecological basis necessary to sustain human civilization.

The MACS framework laid out the biophysical adaptations and spectralizations necessary for economic development. But technologies can only emerge and proliferate if they're selected by social groups. And the selection process depends on the class-power dynamics that characterize a particular society. These dynamics both influence the process of technological selection and are also influenced by it in return. It's precisely here that social concepts like *class*, *institutions*, and *hierarchy* need to be invoked as organizational principles. Kings and pharaohs command more influence than workers and managers. For the most part, the dominant classes, corporations, and governments of the world attempt to calibrate their economies on terms that are favorable to themselves. They modulate energy and financial flows to suppress destabilizing or revolutionary activities by workers, to score higher profits, and to control the political process. The development of finance, much like that of technology, is an emergent social process that unfolds in response to a complex array of economic and political conditions.

It's tempting to look at the nominal domain and view it as a kind of reflective mirror that reveals the inner workings of the real domain, with its spheres of production and biophysical dynamics. In reality, the world of money reveals far more about social relations and social struggles over economic distribution. It's the social domain that mediates and manages the complex relationship between the real domain of production and the nominal domain of money, and changes in the real and nominal domains also affect the evolution of the social domain itself. In our current age, capitalists adapt and structure technological change to dominate the class struggles over distribution unfolding through the various political, economic, military, and cultural domains of human society. Capitalists accumulate capital and increase it as much as possible, in competition with other capitalists. They want to seize anything in the nominal domain that's necessary to organize the world on their own terms in the face of social resistance. Capitalists strive to accumulate profits, revenues, asset valuations, and market capitalizations, to name just a few. And as we'll shortly see, the capitalist system has a tendency toward the concentration of capital in all spheres of the capitalist economy, leading to monopolies, oligopolies, and conglomerates that dominate their respective markets.[3] Conflicts between labor and capital are always there, but not always in the open. The accumulation of capital can also run into roadblocks in the spheres of both production and distribution, and in recent times these roadblocks have caused a period of secular stagnation and slower growth rates across the advanced capitalist world.[4]

Capitalists pursue and implement accumulation strategies in many different ways. Accumulation can manifest as an offshoring strategy, like when Nike shipped jobs to Asia back in the 1980s and 1990s as a way of reducing labor and production costs and raising profits relative to its competitors. Offshoring is directly implicated in one of the central problems of the current international system: the energy-intensive nature of global trade. In a landmark 2012 paper, a group of Australian researchers analyzed the aggregate raw materials exchanged among countries through international trade and introduced the concept of the *material footprint*, defined as the global allocation of used raw material extraction to the final demand of an economy. They concluded that "with every 10 percent increase in gross domestic product, the average national [material footprint] increases by 6 percent."[5] In their view, "Achievements in decoupling in advanced economies are smaller than reported or even nonexistent." They also estimated that roughly 40 percent of all global raw materials are extracted to facilitate the export of goods and services to other nations, which indicates that reducing the international flows of global capital could be a critical strategy in addressing our intensifying ecological crisis.

The drive for financial accumulation can also happen in the context of automation, such as when companies replace call center workers with chatbots, when retail chains replace their cashiers with self-service stations, or when factories replace workers with robotic arms. The economists Daron Acemoglu and Pascual Restrepo defined automation as "the development and adoption of new technologies that enable capital to be substituted for labor in a range of tasks."[6] They're using the term *capital* to mean a factor of production, like factories and machines. Contrary to popular misconceptions, however, automation isn't always about replacing labor with machines. Human beings have been automating productive tasks for millennia and people still have jobs. Technological adaptation and spectralization—in other words, "innovation"—simply introduce new problems that have to be confronted, new challenges that have to be resolved, new tasks that need to be completed, new dynamics that have to be navigated. Precisely because labor is the fundamental social source of production, the ruling classes of any age will always find a way to exploit, manage, and distort it for the purposes of accumulation, regardless of how much automation occurs along the way. In that sense, automation isn't really about eliminating work, but about *changing and transforming the nature of work* so that it conforms to the class-power dynamics of a given historical age.

The rise of artificial intelligence (AI) and software algorithms are the perfect examples of this historical process. AI has generated a great deal of fear about people losing their jobs and silly doom loops about the end of civilization. But the real purpose of AI is to serve as another powerful mechanism of class control, another way for capitalists to sabotage workers and to ensure their class domination over society at large. Capitalists will end up exploiting AI to constrain the wage dynamics of the broader economy, to eliminate potential rivals in the market, to find new ways to punish and control labor, to convince us to buy more useless things we don't need, to get us to click on more online ads and watch more stupid videos, and to make us believe foolish things about the world that align with the financial interests of dominant corporations. AI will be, and has already been, subsumed under the wider social forces and processes of accumulation. For example, generative AI has recently captured the public imagination, and for good reason. It's quite amazing to see AI chatbots communicate like human beings or to see AI systems generate entirely unique videos based on text prompts. But the rapid expansion of the AI revolution is coming with enormous environmental costs. Because they need huge amounts of data, advanced AI models are trained through networks of computers arrayed in server racks stored in massive data centers. These data centers consume vast amounts of electricity. Whereas electricity generation in the United States largely flatlined and stabilized for the past two decades, it's now projected to rapidly increase in the next decade as more and more data centers are constructed for cloud computing, generative AI, and other related computing endeavors. At the end of 2023, U.S. grid planners projected a nearly 5 percent increase in power demand over the next five years alone.[7] The growth of AI risks pushing global civilization over the environmental precipice even faster.

Like virtually all aspects of technological change, automation is first and foremost about social power and class struggle. It's about making labor more obedient and precarious. It's about overcoming energetic constraints and expanding production. It's about seizing new markets and eliminating competition. And it's also about constraining wage growth and changing the wage dynamics of a particular sector or in the economy. In London, cab drivers once earned a decent income because they had to pass a rigorous test, called "The Knowledge," where they revealed their encyclopedic mastery of the city's streets purely from memory. But then Uber came to London in 2012. Uber drivers relied on smartphones to get around, so they had the entire city's road network on the palm of their hands. Since that time, Uber has created jobs for more than 40,000 drivers, but by eliminating the requirement for extensive knowledge that previously justified a large wage premium, it has also produced an economic landscape of much lower incomes among taxi drivers.[8] Uber's basic business model when it opens in a new location, whether London or anywhere else, is to incentivize lots of early drivers to join with high initial compensation rates, and to then gain market share relative to competitors by cutting fare prices.[9] Because of lower fare prices and so many Uber drivers on the road, income almost always falls dramatically for individual drivers over time.

In an ideal world for them, capitalists would absolutely love to make more money by holding production constant, or scaling it back altogether, while at the same time increasing prices for goods and services ad infinitum. That way they can cut back on production costs while simultaneously generating higher revenues because of the higher prices for their goods and services. But if carried too far, this strategy will produce dangerous levels of inflation, which can make societies politically unstable and economically dysfunctional. No one wants to be Weimar Germany in the early 1920s. In the long run, then, political pressure typically forces capitalists into organizing production for the purpose of volume growth—making more commodities, in other words. But producing more runs the risk of generating substantially higher costs unless the production process is properly managed by the capitalist. To control and reduce the resulting financial costs associated with this strategy, capitalists routinely turn to automation.

In the history of capitalism, automation has been one of the major selection vectors for technological change because it activates an arms race as capitalists rush to scale back and reduce labor and *unit* production costs, always aiming to score higher profit rates than their rivals. Because automation facilitates mass production at lower unit costs, it allows dominant corporations to engage in predatory pricing to undercut their rivals. It also gives dominant capital a huge amount of leverage in global trade. The mechanization and automation of modern agriculture is a prime example, as transnational food corporations can sell staple goods like rice and wheat at much lower prices than local farmers in many poor countries.

Although automation has been critical to modern history, it would be wrong to view it as the teleological destiny of capitalism. Automation significantly contributes to the emergence of technological adaptations and spectralizations, but not all new technologies diffuse across the economy right away. There is a lazy assumption, among most economists and the public at large, that capitalism goes hand-in-hand with unfettered technological growth. However, actual capitalists care about technological innovation only to the extent that they can successfully commodify the resulting products and their corresponding intellectual property rights. Technological innovation is useful to capitalists only if new technologies become profitable, but new technologies are often *not* profitable for long periods of time. In many cases, for example, they only diffuse because of state-driven industrial policy that establishes entirely new markets for their use and consumption. Consider the adoption of the catalytic converter, an amazing device that removes toxic chemicals from tailpipe emissions and one of the classic examples of concerted government action creating a market for a new and expensive product. Catalytic converters only became widely adopted by car companies after the Environmental Protection Agency forced the auto industry to change its ways in the 1970s, dragging GM and Ford through brutal court battles that had the effect of both weakening the original regulations and still requiring the car companies to install the converters.[10] The auto industry took a financial hit from the resulting installations, which is why it was never going to adopt the converters without state action, but the regulations also dramatically improved air quality across the United States, especially in urban areas.

Technological innovation can often be an albatross for the class domination of capital. If circumstances arise that make technological changes and innovations unprofitable, then the corporate world is all too willing to fight against their implementation. One need only look at the two-faced nature of the auto industry, publicly supporting efforts to curb global warming on the one hand, while practically delaying the implementation of more fuel-efficient vehicles on the other. When the United States government bailed out the auto industry in the Great Recession, to the tune of $80 billion, the car companies agreed to adopt much more stringent fuel efficiency standards for their future vehicles.[11] But after Donald Trump was elected, he scrapped the agreement at the behest of the very same companies that had appeared so willing to make more fuel-efficient cars only a few years ago. Perhaps their former willingness depended more on sucking up to the government in charge at the time for long enough to stay alive.

Capitalists are indeed skilled at blabbering about efficiency and innovation as they push for financial accumulation and expand the underlying scale of their energy systems. In 1924, the world's major light bulb companies formed a secret cartel called Phoebus, after the Greek god of light. The main purpose of the cartel was to mandate new manufacturing standards that would reduce the planned lifetime of an incandescent light bulb. By the early 1920s, rapid technological development had significantly improved the lifetime of the average light bulb, to the point where many of them could last as long as 2,500 hours. But that turned out to be a problem for the light bulb companies: better-lasting bulbs were hurting sales because consumers could wait a much longer time before buying replacements. To remedy the problem, the companies of the cartel ordered their engineers and manufacturers to produce light bulbs that would only last 1,000 hours. Their public justification for this move was an apparent desire to improve efficiency, but recently discovered internal documents from some of the companies make it apparent that the real reason for the shift was to boost sales.[12]

The formation of the Phoebus cartel has sometimes been called the birth of *planned obsolescence*, which is the process of designing and manufacturing products to ensure their rapid breakdown, thus bringing consumers back to purchase additional versions of those defective products. Planned obsolescence is, in effect, a form of technological sabotage disguised as an apparent innovation. It has become a cornerstone of corporate planning under capitalism, impacting everything from personal vehicles to electronics. For example, Apple became notorious for designing smartphones, laptops, and other products that were engineered to fail within a couple of years.[13] They have faced multiple lawsuits over precisely this issue, but they keep settling out of court and the billions keep flowing in. Apple doesn't care about what it's doing because it costs virtually nothing to pay off disgruntled customers, relative to the obscene prices it charged them for those products in the first place. Because high-tech commodities are designed to fail quickly, they must be continuously reproduced, repackaged, and reshipped so they can be sold to consumers. That process requires large amounts of energy. One of the biggest effects of planned obsolescence is to reinforce the energy-intensive nature of capitalist production and distribution.

Driven by the imperative of accumulation, dominant corporations and monopolies use many strategies to artificially manipulate prices, reduce costs, and control the levers of supply and demand in the economy. For example, corporations will create artificial scarcity and restrict market supply as a time-honored capitalist tradition designed to make goods and services appear to be more exclusive. Before the days of streaming apps, Disney had a habit of randomly pulling its animated classics from the home video market in order to drive up demand, marketing them as being sealed in the "Disney Vault."[14] Dominant meat corporations, like Cargill and Tyson Foods, deliberately purchase very few cows from ranchers and feedlots in order to keep the supply artificially low, thus justifying higher prices for the meat products they sell to restaurants and grocery stores.[15] This strategy was an especially big problem in 2021, right after the first major Covid wave subsided, when cows were readily available in large numbers and the meatpacking industry had plenty of workers to operate the slaughterhouses. But the meat companies jacked up beef prices anyway, simply because they could. This strategy crippled many ranchers and breeders who invested so much time and money into growing their cows, only to find out that their animals were increasingly worthless because of the supply glut that the large meatpackers had artificially engineered.

The class-power struggle behind the process of accumulation is designed to distort supply and demand cycles in favor of elite capital. That's because capitalists operate by setting and establishing their desired parameters in the nominal domain, then altering and adjusting the real biophysical sphere of production in accordance with those goals. Capitalists want to hit certain profit rates and revenue targets, and they'll change whatever they need to change about the biophysical world in order to reach those goals. They'll fire half the workforce of the company, as Elon Musk did when he took over Twitter (now X). They'll cut back on production and raise prices if they think they can get away with it. They'll invest in new technologies that they believe will give them an edge over competitors, which is what Microsoft did when it poured $10 billion into Open AI, the maker of ChatGPT, as a way of trying to sprint past Google.[16] They'll ship jobs overseas and automate whatever they can. They'll do some other things, which often escape the immediate attention of consumers. Corporations use *shrinkflation*, the size reduction of commodities while leaving prices unchanged, to lower supply costs and to therefore score higher profits. For example, back in 2016 Mondelez International reduced its 170-gram Toblerone chocolate bar to 150 grams, but sold the item for the same price.[17] That move came on the heels of another size reduction just a few years earlier. Companies also use zone pricing for gasoline and other products, a strategy where consumers are charged different prices for the same commodity depending on where it's sold in a regional market. This strategy is notorious for producing high levels of local price dispersion and variation for common staple foods like milk, meat, and plenty of other food and drink items.[18] Research has shown that, in many cases, most of these variations cannot be explained by store heterogeneity, like the fact that you might expect food sold at airport convenience stores to be more expensive than elsewhere.[19] There are enormous price variations among goods sold at grocery stores that are located in the same neighborhoods and small towns. Other strategies of corporate accumulation include price matching by companies to seize market share from competitors. For example, Target will price-match any of its products that are being sold by Amazon or other online competitors at a cheaper price.

In truth, the entire architecture of capital accumulation is implicated in this saga. Even tax laws and accounting rules, among other regulations, are subsumed under the class-power dynamics of modern capitalism, as corporations and the wealthy pressure the state to set rules and regulations that are favorable to financial accumulation for dominant capitalists. An egregious case of corporate tax plundering, for example, is the deduction given for accelerated depreciation in the United States. This provision allows companies to claim a faster rate of depreciation on their capital equipment in the early years of an asset's life. Accelerated depreciation stands in contrast to the straight-line method, in which companies evenly spread out the cost of the asset across its expected lifetime. In 2011, the U.S. Treasury Department estimated that accelerated depreciation would cost the federal government $270 billion in lost tax revenue over the next five years alone, a staggering sum that still probably underestimates the real cost.[20] The Trump tax law passed in 2017 inflamed the situation even further. That law introduced the notion of "bonus" depreciation, which allowed companies to deduct *the full value* of an asset right away, in the very first year it was purchased. A 2022 analysis concluded that accelerated depreciation under the Trump tax law had saved twenty-three major corporations an astounding combined total of $50 billion from 2018 to 2021.[21] For example, Verizon obtained roughly $5 billion in federal tax breaks from accelerated depreciation alone during that period, an amount that was almost half of the total federal tax breaks it received.

In the class-power dynamics that prevail in the capitalist system, the entire natural world and the biophysical sphere of production are subjected to the whims and machinations of a corrupt parasitic class that happens to be in a position of power. One of the more pernicious forms of accumulation under modern capitalism is the process of monopolization, the corporate drive to dominate a particular market by eliminating competitors. From Standard Oil to AT&T, capitalism has a decisive tendency to create monopolies. Amazon is a prime example in recent times; it got to be so dominant chiefly through predatory pricing and killer acquisitions, both strategies specifically intended to eliminate its competitors.[22] Over the years, Amazon has seized companies like Quidsi, which operated a popular baby products website, and Zappos, a major online shoe platform. The basic goal was always the same: find potential competitors early, then gobble them up as soon as possible. And this entire drive for accumulation was fundamentally based on Amazon's general strategy of cost reduction. Amazon is infamous for bullying third-party vendors into accepting lower sales prices on its platform, thus driving up demand and gaining market share relative to other online retailers. Amazon also exploits a high-turnover labor force in massive distribution warehouses where workplace injuries and deaths are common. Amazon warehouse employees are so ruthlessly overworked that they have little time to go to the bathroom and many of them consume painkillers from workplace dispensers to deal with the pain they've sustained on the job. Amazon's drivers are also pressured to meet delivery schedules with tight deadlines, leading to accidents and dangerous behavior on the road. The result is an energy-guzzling behemoth that has massive strategic influence over the American economy. For 2019, Amazon reported that it was responsible for 51 million metric tons of carbon dioxide emissions, the equivalent of 13 coal power plants running for a year and a 15 percent rise from the previous year. Even this figure, however, is likely a severe underestimation of the real number, given that corporate accounts on emissions and energy use are based on flawed methodologies.[23]

One of the great pitfalls of monopoly power is the suppression of competition and innovation, especially if that innovation is not compatible with existing business models. Consider that Bell Labs, the elite private research division of AT&T in the twentieth century, invented magnetic tape recording and even created a prototype answering machine in the 1930s before shelving the products because executives feared that people would avoid using telephones if they believed that their conversations could be recorded.[24] Magnetic tape recording wouldn't become available until 1962 through the audio cassette, an arrival that was due to companies outside the United States. Bell Labs also invented and set aside early versions of mobile phones, DSL, fax machines, speaker phones, and fiber optics.[25] The demise of the AT&T monopoly had a major positive impact on the telecommunications industry and on technological innovation in general. In 1968, the Federal Communications Commission issued new rules that allowed users to connect devices made by external manufacturers to the network of AT&T. The decision had an immediate effect, leading to "rapid innovation in end-user devices, such as answering machines, fax devices and…modems."[26] In the 1980s, the federal government broke AT&T into several companies, although those companies gradually got back together through multiple rounds of mergers and acquisitions, as part of a broader intensifying wave of corporate monopolization in the American economy.

Monopolization as an economic process isn't always about producing one dominant company. Sometimes it can produce a cartel that implicates many companies colluding together, as we've already seen. Several major U.S. music companies in the late 1990s reached a price-fixing agreement for CD sales, preventing retailers from offering discounts and avoiding competitive price offers.[27] From 1999 to 2006, global producers of liquid crystal display (LCD) panels, which are used on everything from television screens to computer monitors, were in collusion together to fix prices.[28] The companies involved included the likes of Toshiba and LG Display. These companies would exchange information on production plans, shipping schedules, and pricing strategies. They were aware that their actions violated competition laws, and they took great pains to avoid public scrutiny by holding their regular meetings in hotel rooms. Once lower-level employees replaced the top brass at the monthly meetings, the venues shifted to public restaurants. Total fines against the conspirators, from multiple international investigations, totaled well over $1 billion. Almost no industry in the world is spared from the cartel syndrome; it's one of the most addictive imperatives of modern capitalism. For another example, a major global auto parts cartel in the 2000s involved over 100 companies fixing prices for car parts sold to companies like Toyota and Honda. A study from 2019 estimated that the cartel's activities affected anywhere from $3.2 trillion to $5 trillion of international commerce, with injuries and overcharges sitting somewhere between $600 billion and $1 trillion.[29]

It stands to reason that if financial accumulation is the guiding social principle underlying technological change in the capitalist economy, then overthrowing and sabotaging the current rules of capital accumulation should be a central priority in our efforts to remake and reconstitute the biophysical relationship we have with the natural world. The central problem is that there are no limits on this process, and therefore no limits on the exploitation and expropriation that capital can inflict on humanity and the planet. For one potential solution, governments can consider imposing maximum caps on the parameters of financial accumulation, such as by setting aggressive limits on the market capitalization that firms can reach and by capping the total amount of wealth that an individual may own at a given point in time. Any amount of wealth exceeding these thresholds would automatically become public property, belonging to the people or the state. These caps could also be dynamic and change over time, reflecting the changing characteristics of society. And by virtue of being dynamic thresholds, they would still allow various forms of competition and innovation to occur, thus giving people incentives to start new businesses, create new inventions, and market new products. However, by sharply constraining the possibilities for accumulation, these thresholds would also dissuade the aggressive drive for financial accumulation that currently dominates the capitalist order.

In summary, the nominal domain largely reflects social and power struggles over the distribution of economic resources, not objective factors about productivity or efficiency. At the highest levels of social power, these struggles are driven by a process of accumulation in which capitalists try to seize more wealth and capital to outcompete their rivals, and to control the labor power of the working classes. The class-power struggles over distribution nevertheless fundamentally depend on the underlying labor relations and the productive methods of society, as people can only squabble over the wealth available in society after it has been created. Furthermore, the class-power dynamics of society have major consequences for the biophysical and energetic basis of the economy, which implies that institutional or revolutionary changes to those social dynamics can profoundly change the biophysical dynamics of the economy itself. Combined with the prior arguments about the nature of technological innovation in modern times, there's a compelling case that the best way of handling the bionomic disruption is to change how society works, not to blindly focus on technological innovation and hope for the best. We should focus on changing how governments and corporations are organized, on changing how power is wielded and distributed within existing social institutions, and on creating new local and global institutions that can properly and effectively address the massive challenges we face. Technological innovation can be part of the solution, but it should not be seen as the central plank, and it certainly cannot get the job done all by itself.

The State Dynamics of Energy Transitions

Technological change is driven and organized by class-dependent and hierarchical institutions, and none is more important in this regard than the state. Consider the development of the internet, one of the most revolutionary changes in human history. The technical foundations of the internet were established in the 1960s and 1970s through extensive government funding and coordination, especially from the Advanced Research Projects Agency (ARPA) at the U.S. Department of Defense.[30] The Eisenhower administration formed ARPA, along with a slew of other state-led initiatives and policies designed to enhance scientific progress, in response to the Soviet Union launching Sputnik in 1957. The internet, an obsessive and indispensable part of our lives, was a byproduct of the geopolitical rivalries of the Cold War. The Global Positioning System (GPS) was also developed and funded by the U.S. Department of Defense; it has become indispensable for communication and transportation needs over much of the world.[31] Critics of state-led industrial policy will point out that government investments often fail, which is certainly true. However, the same thing is true of private sector investments. Conservatives were quick to lambast the Obama administration for funding the now-defunct solar company Solyndra back in 2010, but they ignored the nearly $500 million loan that the government gave in the same year to the little-known company called Tesla.[32] That low-interest loan was absolutely crucial for Tesla as it allowed the company to finish building its Model S factory in California, and the rest is history.

In modern times, the role of state funding and organizational capacity on technological innovation has been absolutely critical. Consider that about 25 percent of the Bell Labs semiconductor research budget over the period 1949 to 1958 was funded by defense contracts.[33] In 1959, a congressional committee estimated that roughly 85 percent of R&D costs in U.S. electronics were paid for by the federal government.[34] A 2003 estimate noted that government funding was "directly or indirectly" responsible for "40 to 45 percent of all industrial R&D in the semiconductor industry between the late 1950s and early 1970s."[35] For another example, Taiwan's government provided nearly 50 percent of the initial capital for the 1987 founding of TSMC, now the world's dominant semiconductor manufacturer.[36] The United States government gave over $10 billion in funding to pharmaceutical companies like Moderna and Johnson & Johnson as part of Operation Warp Speed, the name of the plan to rapidly develop vaccines and treatments for the coronavirus. Those vaccines led to surging sales and profits for these companies in 2021 and 2022. As economist Marianna Mazzucato has pointed out, the state is usually first to invest in highly risky and uncertain ventures.[37] The private sector comes in later, after the initial uncertainties have been resolved, to further refine the technologies developed by the state. There would be no SpaceX without NASA.

Governments don't just stand around in peacetime handing out money and investing in future technologies. They also go to war against each other, and war has been absolutely instrumental in the process of technological change and financial accumulation. As they try to gain a competitive advantage against their enemies, governments invest huge sums of money in wartime for the production of weapons and the development of new technologies. The most notable case of conflict-driven technological innovation is clearly the Second World War, the fundamental technological singularity of modern times. The vast majority of all advanced technologies used by humanity today, from smartphones and computers to lasers and transistors, can all be directly or indirectly traced back to developments in that war. Let's consider transistors, tiny semiconductors that amplify and control electrical signals. They're the foundational component of the integrated circuits in computer chips. A typical chip may have hundreds of millions of transistors. Transistors are incredibly useful because they can rapidly alter and amplify incoming electric currents, exactly the physical abilities that are necessary for tiny devices to model the vast numbers of logical and mathematical instructions used in modern computing. Transistors underlie the operations of computers, tablets, smartphones and just about all other digital devices that collectively hook and ensnare the masses nowadays. Perhaps the most revolutionary technology in human history, transistors were invented in 1947 by three American physicists working for Bell Labs. The very first transistor used two gold foil contacts sitting on a germanium crystal. Germanium had been identified as a promising semiconductor during the Second World War, when Western governments were conducting research into radar technology. And after their initial discovery, the diffusion of transistors also happened in large part thanks to the U.S. government, which launched an antitrust campaign against AT&T and forced it to relinquish all transistor-related patents, thus paving the way for the era of the personal computer.

At critical moments in history, the state can also serve as a powerful conduit for dismantling older energy systems and replacing them with newer systems. As long as the energy basis and the class-power relations of a society are fixed in place, then it's extremely difficult to carry out major energy transitions. That's because the coronets of any given economy are adapted for specific social purposes, and those purposes are often decided by the rich and powerful as a way to extend and reinforce their wealth and power. An energy transition is the rapid shift in the prime movers, contingent devices, or primary fuel supplies used by an economy to power its activities. It's certainly a common view among energy scholars like Vaclav Smil that energy transitions are slow and gradual, not radical and discontinuous. According to Smil's reconstructions, coal reached 5 percent of the global primary energy supply around 1840, 15 percent by 1865, 25 percent by 1875, and 50 percent by 1900.[38] Meanwhile, oil's share of the global market reached 5 percent in 1915 and followed similar time intervals to the ones cited above for coal. Natural gas held 5 percent of the global market by 1930 and 25 percent of it after fifty-five years, "taking significantly longer to reach that share than coal or oil."[39]

From time to time, however, every social system is faced with a crisis, a revolution, or a novel opportunity, any of which may be driven by sudden exogenous factors outside the system, by endogenous ones building up over time, or by both acting concurrently in time and space. These crises or discoveries can provide golden opportunities for societies to implement radical energy transitions on quick time frames, as the energy scholar Benjamin Sovacool has observed.[40] Sovacool writes: "Transitions to newer, cleaner energy systems such as sources of renewable electricity, or electric vehicles, often require significant shifts not only in technology, but in political regulations, tariffs, and pricing regimes." The transition to nuclear power in France is a canonical example. Following the oil crises of the 1970s, France made the decision to use nuclear power for electricity production. France constructed fifty-six reactors from 1974 to 1989. Nuclear power went from 4 percent of the country's electricity supply in 1970 to almost 40 percent by 1982, growing nearly tenfold in just twelve years. France was able to institute such a massive change in its electric coronets largely because of a national government that centralized decision-making and wielded enormous power over the economy, establishing the nationalized utility company EDF along the way. Of course, as we saw in chapter 7, later periods of underinvestment and lack of state capacity have contributed to deteriorating conditions for the French nuclear power industry. But the central point is simply about the initial transition and what was required to make it happen.

The French response to the crises of the 1970s is certainly not an isolated example. After the Netherlands discovered a massive natural gas field at Groningen in 1959, it implemented a quick transition away from an economy powered by oil and coal toward an economy powered by natural gas. In the year of the Groningen discovery, coal provided roughly 55 percent of the Dutch primary energy supply, followed by crude oil at 43 percent; natural gas had less than 2 percent of the total share. But then things began to change. Natural gas supplied 5 percent of the primary energy supply by December 1965, one year after gas deliveries started from Groningen. Just six years later, the natural gas share of total Dutch energy supply had reached an astonishing 50 percent. The government implemented this transition by abandoning "all coal mining in the Limburg province within a decade, doing away with some 75,000 mining-related jobs impacting more than 200,000 people."[41] To compensate for such a risky gamble, the Dutch government issued subsidies for new companies, started training programs for laid-off miners, and relocated state industries away from Amsterdam to other urban and rural regions impacted by the mine closures. The result was a rapid energy transition that not only failed to kill the Dutch economy, but actually laid the basis for much of its future success. Even as late as 2010, natural gas provided 45 percent of Dutch primary energy supply, the largest single source.

The Dutch transition toward natural gas is a powerful reminder that any radical energy transition needs to be properly managed by the state so as to avoid economic hardships for the masses. Too often in public discourse dominated by corporate media or elite academics, there's a fawning and often silly obsession with the market incentives that the government needs to establish to eliminate our addiction to fossil fuels. It's in this context that subjects like carbon taxes and cap-and-trade schemes come up. What these discussions often ignore is that such proposals are likely to be highly destabilizing for the working classes unless powerful state interventions dilute their more harmful effects. For example, if a government imposes severe carbon taxes on the corporate world for emitting greenhouse gases, then corporations will simply pass on those extra costs to their consumers, leading to higher commodity and energy prices. Governments therefore cannot simply set up market incentives and hope for the best; they must actively redirect and reorganize large parts of the economy specifically to achieve the major goals they've established. This means they must assert and impose more direct levers of control over economic activity.

The central theme here is that technological change does not have to be the only causal vector for implementing an ambitious energy transition. Sudden transitions can indeed occur because of state-led initiatives or major sociopolitical changes. And these transitions don't have to be gradual. In fact, precisely when classes and states find themselves in critical moments full of revolutionary potential, energy transitions can proceed on very large scales and at incredibly rapid speeds. On a global level, Smil's point merely shows that different global institutions and arrangements would be needed to carry out a radical break with our current energy systems, not that such a radical break is impossible to achieve. Energy transitions take a long time to sweep the world in large part because the global system under capitalism is designed to enrich certain nations while impoverishing others. By remaking global institutions into more democratic and cooperative political bodies we can turbocharge the transition away from fossil capitalism and toward a new era of technological development that's compatible with the prosperity of civilization and the stability of the global ecosphere. That's going to be the main message of the last three chapters, but there's a major test that the ecodynamic synthesis first needs to pass before we can reach its prescriptive stage, and that's explaining the emergence of the Industrial Revolution, because only by properly understanding how we reached our current mess can we ever hope to get out of it.

CHAPTER 9

The Industrialization of Britain: A Case Study

Everything about the Industrial Revolution is debated. When did it start and when did it end? Was there one industrial revolution or multiple industrial revolutions stretching across several centuries? Many historians occasionally speak of the First Industrial Revolution from about 1750 to the middle of the nineteenth century and the Second Industrial Revolution from about 1870 to the outbreak of the First World War. In recent times, some writers have even talked about a "Third" Industrial Revolution related to the rise of the internet, digital technology, and renewable energy.[1] Many scholars prefer to emphasize the "Great Acceleration" that occurred after the Second World War, in which the global AFR witnessed a massive explosion, the human population skyrocketed, and the impact of civilization on the entire biosphere reached unprecedented heights. The main aim of this chapter is to explain the emergence of the First Industrial Revolution in Britain, since all later waves of technological development and economic growth in modern times flowed from that initial spark.

Other questions about the history of industrialization abound.

Should industrialization be seen as a radical, discontinuous transition to a new kind of economy or was it a more gradual and continuous transition from the medieval economy to the early stages of modern capitalism? In other words, was it the Industrial *Revolution* or the Industrial *Evolution*? How dependent was the Industrial Revolution on slavery and the Atlantic trade system? Should the Industrial Revolution be understood as a largely internal and autonomous process unfolding within Western Europe or as a broader phenomenon in which European historical dynamics were fundamentally entangled with external developments outside of Europe? Did colonialism and imperialism predate and facilitate the Industrial Revolution or was it the other way around, with the rise of industrialization making possible imperialism and colonialism? Were constitutional and institutional changes more responsible for industrialization, or do technological innovations and bold entrepreneurs deserve more of the credit? These are just some of the important questions and controversies swirling around the Industrial Revolution, and it's hopeless to pretend that we can answer all of them here. But what we can do is provide a comprehensive explanation of the Industrial Revolution that provides useful context, if not necessarily strict answers, for all of these questions.

The Ecodynamic Synthesis and the Industrial Revolution

In *The Unbound Prometheus* (1969), the historian David Landes placed the Industrial Revolution in a wider historical context, arguing that it was part of a larger wave of modernization. According to Landes:

> Industrialization is . . . at the heart of a larger, more complex process often designated *modernization*. This is that combination of changes—in the mode of production and government, in the social and institutional order, in the corpus of knowledge, and in attitudes and values—that makes it possible for a society

to hold its own in the twentieth century; that is, to compete on even terms in the generation of material and cultural wealth, to sustain its independence, and to promote and to accommodate further change. Modernization comprises such developments as urbanization…the establishment of an effective, fairly centralized bureaucratic government; the creation of an educational system capable of training and socializing the children of the society to a level compatible with their capacities and best contemporary knowledge; and of course, the acquisition of the ability and means to use [updated] technology.[2]

Landes became famous for his account of the technological leaps and bounds that made the Industrial Revolution possible, especially from the late nineteenth century onward. But his basic causal explanation is naïve because, in typical fashion for cultural and idealist theories, it implicitly takes the modernization process as a given *deus ex machina* that just comes out of nowhere. Many aspects of modernization that Landes mentions in the passage above, such as changes in the mode of production or changes to the corpus of knowledge, were largely the products of industrialization as a historical process. It was precisely the changes to the underlying productive methods and economic forces of society that generated further changes in culture, knowledge, values, and ideas. And those deep underlying changes to the coronets of industrializing economies started off, in large part, because of unexpected ecological shifts and ruptures that human beings did not control. In this sense, only materialist theories can properly ground broad historical changes to the social metabolism in the wider architecture of the natural world. As I explained in chapter 3, cultural theories of economic development are plagued by too much idealism, and thus have no concrete sense of causation or causal context. The reality is that behavioral changes among people, including the behavioral shifts of the inventors and investors of the Industrial Revolution, can be best explained by invoking economic incentives, ecological pressures, and the broader class dynamics that mediate the relationship between those incentives and pressures.

Marriage patterns in premodern Europe provide a good example. Historians have noticed that women in northwestern Europe during the Late Middle Ages waited longer to get married than women in other parts of Eurasia and Africa. Whereas women in these latter parts of the world would often get married in their teens, many women in Western Europe waited for marriage until their twenties. The end result of this delay was that Western European women had, on average, fewer kids. A smaller average family implies that goods and services are spread out over comparably fewer people, thereby raising living standards. But what explains this difference in marriage patterns between Western Europe and other regions around the world? A plausible explanation is the relatively high wages that developed in several northwestern European countries after the Black Death.[3] The extreme labor scarcity caused by the huge death toll in Western Europe prompted many women to continue working longer because they no longer had the same economic pressures and incentives to get married. So, what seems like a *cultural* explanation on the surface can be exposed as an *economic* phenomenon upon further examination. Cultural explanations for the Industrial Revolution also focus on supply-side constraints but deal very little with the demand side of the story. For example, greater scientific knowledge can certainly contribute to conversional spectralization, such as when people learned that the atmosphere actually has weight and can apply pressure. This discovery in the seventeenth century set the framework for the development of atmospheric steam engines in the eighteenth century. But still, new technologies can only arise if there's some kind of demand for their existence, and that part of the story can only be understood once we examine the economic incentives that are driving people to make their decisions.

Running on a semi-parallel track to the cultural school, liberal and institutionalist perspectives emphasize the emergence of limited parliamentary governments as *the* important step on the road to capitalism. Many liberal historians argue that the Glorious Revolution of 1688 made the Industrial Revolution possible because it limited royal authority, consolidated the power of the English Parliament, and secured private property rights. I've already discussed some of the foundational problems with neoclassical institutional theories, but I can mention a few more in the context of the Industrial Revolution. First, studies of banking and interest rates "fail to detect any structural break after 1688," meaning that no financial indicators can be found to support the idea that the investment climate improved noticeably.[4] Second, property rights in France, and even in China, were at least as secure as those in England during the eighteenth century. In fact, France might have had too much in the way of property rights, as evidenced by the example of profitable irrigation projects in Provence, a territory in southeastern France. These projects were not completed because property owners opposed the enclosure of their land for public use and did not want to construct any canals or turnpikes.[5] They only started after the French Revolution limited local liberties and concentrated more power in the national assembly. Meanwhile, the British government had no problem forcing property owners to give up certain rights. Third, taxes were higher in England compared to France during the eighteenth century.[6] All in all, institutional and constitutional theories cannot adequately explain the scale of the energetic and technological changes that happened in the Industrial Revolution.

In his major 2009 work, *The British Industrial Revolution in Global Perspective*, historian Robert Allen emphasized the importance of cheap energy and high labor costs in making England a unique environment for investment in new technologies. With a cheap and abundant source of energy suddenly everywhere, inventors and businesses decided to invest more time and money in labor-saving technologies. As Allen puts it, "The success of the British economy was...due to long-haired sheep, cheap coal, and the imperial foreign policy that secured a rising volume of trade."[7] In 1800, coal output from British mines was 66 times higher than what British mines could produce in 1560. This enormous supply of cheap energy depended on technological innovations, but also on plain good fortune: England just happened to be sitting on massive deposits of coal that were strategically located and easy to exploit. The influx of silver from the New World, along with various demographic and political changes, led to wage and price inflation throughout Europe. This inflationary pressure was especially pronounced in northwestern Europe because of its high death toll from the Black Death and other recurring plagues. In the Late Middle Ages, for example, wages were roughly equivalent in all major European cities. By the year 1600, however, workers in London and Amsterdam had pulled well ahead of their counterparts in Europe and Asia. English and Dutch laborers made about 8 grams of silver a day in 1600 whereas European laborers in other major cities made somewhere between 4 and 5 grams of silver a day.[8] This wage inflation continued unabated in London, to the point where London workers were making roughly 15 grams of silver a day by 1800. Contrast these earnings with the 9 grams of silver a day in Amsterdam and the 4 grams of silver a day for workers in Beijing at the same time.[9]

The relatively high wages in Britain led to a major push for developing labor-saving technologies. Capitalists and industrialists wanted to automate the production process because they could save a lot of money on labor costs if they could get machines to do what human beings once did. But this logic only worked well in Britain because wages there had reached such comparably high levels. Other parts of the world had no equivalent incentive to make the huge upfront investment costs for technological innovation because rich elites in those countries could always find a vast pool of cheap labor to perform the necessary tasks of civilization. This is one of the most important reasons why England invented practical steam engines and Song China did not. In China, cheap labor was readily available and so there was no great incentive to develop complex machinery. The situation was completely different in England. The steam engine was first used to lift water and coal out of mines. Following Thomas Newcomen's engine in the early eighteenth century, many inventors went on to make several marginal improvements to its fuel efficiency and power output. The engines developed by James Watt, for example, more than doubled the efficiency of the Newcomen engines. But these changes were all expensive and could only have succeeded with a regular supply of funding from willing investors, who wanted to invest in new technologies as a way of bypassing high labor costs. The owner of a coal mine in nineteenth-century England could have hired people and bought horses to extract water and coal out of his mine. But getting a steam engine to do the job faster was far more economical in the long run, and so English capitalists used steam engines more and more to replace much of the work that was previously done by miners.

We now have all the necessary pieces of the ecodynamic synthesis to coherently demonstrate how and why industrial capitalism first emerged in England, and not in places like Spain or Sweden. The ecological conditions that led to the development of capitalism spanned everything from ancient and gradual mechanisms to sudden and traumatic convulsions, including a moderate warming process in the Middle Ages that boosted European agricultural productivity and helped the continent to triple its population in about two centuries. But the Black Death, which first struck Europe in 1347, was easily the decisive ecological primer on the road to capitalism and industrialization, because it induced radical modal shifts in the economies of Western Europe, which gradually culminated in a series of conversional spectralizations that led to an unprecedented expansion in the AFR of the entire region, but especially that of the English economy. Making this claim does not commit anyone to the more radical notion that the Black Death was *exclusively* responsible for the rise of capitalism and industrialization. The simple fact is that the rise of industrialization in England, and its later diffusion across the world, depended on the temporal and spatial convergence of many complex ecological, political, economic, and cultural causes. Nevertheless, one can still recognize the fundamental importance of the Black Death among these convergent causal factors without being naïvely reductionist. In 2022, the historian James Belich published *The World the Plague Made*, perhaps the most comprehensive analysis of the impact the Black Death had on the future economic development of Europe. Belich writes: "The Black Death fundamentally transformed…the relationship between labor and capital, creating a strong incentive to replace the former with the latter where possible."[10] Most of the revolutionary technology used in Europe during the Middle Ages came before the Black Death, apart from the Gutenberg press. The long series of plagues starting with the Black Death, however, provided extra incentives to use and to proliferate labor-saving technologies. As Belich puts it, "There were few entirely new inventions, but rather a sharply increased uptake of three existing inanimate sources of energy: water power, wind power, and gunpowder."[11] Before 1350, water-powered mills were typically used to grind grain and produce flour, which could then be used to make bread and other food products. But after 1350, water mills not only pounded a higher fraction of grain products, they increasingly shifted to being used for other industrial tasks, such as sawing timber and powering hammers. These initial modal adaptations eventually overcame previous energetic constraints on economic growth because England, in particular, had large concentrations of energy-dense natural resources like coal and relatively high labor wages that encouraged investments in automation. In addition, massive shipbuilding programs and high rates of urbanization induced raw material scarcities in domestic markets and persuaded many English entrepreneurs and governments to heavily invest in the expansion of industrial capacity. Furthermore, the discovery of two strategically exposed continents yielded vast new lands for mining and agricultural production, feeding an imperialist frenzy to seize forced slave labor from Africa and other regions of the world as part of the drive to minimize the costs associated with the production of raw materials for industrial manufacturing.

The growth of European industrial production after the Black

Death is truly remarkable. The raw number of English industrial mills more than tripled from the years 1300 to 1540, going from 600 to roughly 2,000. On a per capita basis that rise represented a sixfold increase because of the demographic hits that England sustained from numerous plague outbreaks.[12] The Black Death was also directly implicated in the rise of the silk industry in the Italian city of Bologna, which became Europe's first proto-industrial hub. By 1371, Bologna was using "12 water-powered silk spinning mills, and the number increased rapidly thereafter to create Bologna's main industry."[13] This elaborate system of silk production relied on a huge underground network of tunnels that moved water around the city, known as *chiaviche*. The operation of this vast network required several hydraulic innovations and remained a closely guarded secret until the middle of the fifteenth century.[14] Bologna's silk industry became famous for producing a thicker kind of silk yarn, known as *organzine* in English, which was then exported to other parts of Italy for processing into finished products and clothing. In the twelfth century, Bologna relied on about seventy grain mills to feed its population; by 1393, it only needed twenty grain mills for the same task, while sixteen mills were used for spinning silk and twenty-one others were deployed by industries like wool, paper, and metallurgy.[15]

The Black Death also had a strong impact on iron production. Before 1350, iron production was largely confined to small bloomeries that generated about "two or three tons of metal a year." After the Black Death, there was a transition toward larger blast furnaces that could produce about "40 or 50 tons a year." Blast furnaces reduced the number of smiths required for executing and managing the smelting process from 30 or 40 to just two or three in some cases.[16] It's been estimated that European iron output "tripled or quadrupled" from 1400 to 1525.[17] The continent's leading exporter, Sweden, experienced a sixfold increase in total iron production from 1340 to 1539, and the per capita figures are almost certainly much higher.[18] Indeed, Sweden became a major supplier of iron to England starting in the seventeenth century and would retain that role throughout much of the eighteenth century as well.

My basic thesis is that the Industrial Revolution in England came down to four major things: sheep, coal, trade, and war. That's an exaggeration, but only a slight one. First came the ecological primer: the Black Death caused a huge population decline, especially in England, which had one of the worst per capita death tolls in Europe. Fewer peasants were available to cultivate crops, so the size of cultivated lands fell dramatically. Lords and peasants converted more land to pasture instead. The rise of pastoralism was a modal adaptation to a severe labor scarcity, since it takes fewer people to herd sheep than to till land to manage crops. Gradual improvements in grass and soil quality yielded higher-caliber wool that was used for the production of textiles and clothing. England's woolen cloth industry became more and more powerful, eventually displacing the Italians and the Flemish as the dominant textile manufacturing center in all of Europe by the seventeenth century. The "new draperies" became responsible for a huge share of London's exports, and they attracted more workers to the city's burgeoning textile industry. In the meantime, England's trade policies and its strategic rivalries with other European nations sustained a growing boom in the construction of warships and merchant ships. A large share of London's workers became employed in the shipping sector, building ships and handling maritime trade. But war and imperialism had another major effect on industrialization in the eighteenth and nineteenth centuries: they led to a massive boom in government debt that drove structural changes in the investment patterns of the English economy. Many investors took the substantial profits they derived from selling government bonds and invested that money into the emerging industries and sectors that would come to revolutionize the country. And by the time of the Second Industrial Revolution in the nineteenth century, British imperialism had amassed large colonial territories that provided critical raw materials necessary for Britain's domination of global trade.

War, trade, and textiles were the dominant factors behind the spectacular demographic growth of London in the sixteenth and the seventeenth centuries. But the enclosures and privatizations of previously common lands, ruthlessly turbocharged by aristocratic forces in Parliament during the seventeenth and eighteenth centuries, also contributed to the emergence of an urban proletariat as poor peasants were violently displaced from their traditional lands and communities. The rapid growth of London then spurred extra demand for timber, because households had to burn wood to stay warm, and the government needed wood to build warships, among other reasons. The rapacious extraction of timber caused local deforestation crises in much of England in the seventeenth century, sending timber prices soaring. Coal then emerged as a ready substitute for declining timber stocks, and the northern parts of England were full of it. We see here another example of an ecological scarcity inducing a shift in the resource base and modal adaptations of an economy. Growing demand for coal led to intensive mining and the search for methods to optimize the extraction process. That's why the first steam engines were invented: to extract water out of coal mines. Steam technology then experienced a series of spectralizations throughout the eighteenth century with the addition of condensers that improved thermodynamic efficiency, new gear configurations that allowed the engines to generate rotary motion and to power machines in factories, and finally the transition from using low-pressure steam to high-pressure steam as the driving motive force, which was the breakthrough moment of the entire Industrial Revolution. Steam engines were already prime movers by the late eighteenth century and became dominant in the British economy by 1870, vastly outpacing the exergy capacity of other devices. The accelerating diffusion of steam power revolutionized all other industries and economic sectors, from textiles to transportation, leading to a massive expansion in the energy scale of the British economy. That, in a nutshell, is how and why the Industrial Revolution first happened in England.

It would be remiss to discuss the causal dynamics of the Industrial Revolution without mentioning anything about agriculture and transportation. Gradual improvements in agricultural productivity, for both crops and livestock, facilitated the expansion of urbanization and were thus critical in shaping the contours of the modern economy. It's been estimated that the milk produced per cow in England rose from roughly "100 gallons per year in 1300 to 380 gallons in 1800."[19] Likewise, the fleece weight from a typical sheep increased from 1.5 pounds to 3.5 pounds. Wheat yields doubled from ten bushels per acre in 1300 to about twenty bushels per acre in 1700. Agricultural productivity increased for several reasons, such as farmers making genetic improvements to their crops and animals as well as the improvements in land and soil quality that resulted from certain enclosures. Advances in transportation also boosted the process of industrialization because it became cheaper to move goods around, thus expanding the available markets for new commodities produced in homes and factories. The construction of new canals and improved roads made it far easier to connect the different regions of the English economy. In particular, Parliament authorized the creation of numerous turnpikes, nonprofit trusts that collected fees from road users and spent much of that revenue on improving the quality of the roads. Before the rise of turnpikes, most English roads had deep ruts and potholes that made it difficult to transport goods on the ground, thus constraining the level of integration between regional markets.[20] By the nineteenth century, England's efficient road system had become a vital lifeline for virtually all businesses and households in the country.

The Adaptations and Spectralizations of the English Economy

Fossil fuels were the keystone resources of the Industrial Revolution because their chemical bonds store enormous amounts of energy. Consider that burning a kilogram of wood releases about 15 megajoules while burning a kilogram of coal releases about 30 megajoules.[21] And burning a kilogram of natural gas yields about 40 megajoules while burning a kilogram of gasoline yields about 43 megajoules. But it's important to note that a given mass of coal occupies roughly a quarter of the volume as the same mass of wood, so the *volumetric* energy density of coal is roughly *eight to ten times higher* than that of wood, an enormous difference. These are all average estimates, and the real-world energy performance of any given quantity of timber or coal will depend on many factors. Nevertheless, they accurately capture the fundamental truth that fossil fuels contain far higher energy densities than sources of energy like renewables or biomass. In the nineteenth century, a few locomotives in the United States actually operated by burning wood instead of coal, but they could never successfully compete against coal-based locomotives because they couldn't store the vast amounts of wood necessary for traveling long distances, a consequence of timber's relatively low volumetric energy density. Any society that can successfully transition to burning coal over wood for most of its domestic needs will necessarily experience a massive expansion in its energy scale, and that's exactly what happened to England.

Before England could shift to using coal, however, there had to be a reason to induce the change. The fundamental reason for that change was the spectacular growth of London from 1500 to 1700. And the main reason why London grew so rapidly is because of England's modal shift toward pastoralism after the Black Death. As millions of acres of arable land were converted to pasture, a series of gradual improvements in land quality and nutrition doubled the typical fleece weight from the fourteenth century to the seventeenth. In the Late Middle Ages, English and Dutch manufacturers started producing a series of enhanced woolen products known as the "new draperies," which were basically light worsteds. Unlike woolens, which are fuzzy and rough, worsteds are based on high-quality wool and are strong, fine, and smooth. These clothing products usually followed the patterns of Italian fabrics, but they eventually became so successful that they "drove Italian producers out of business in the seventeenth century." To boost its domestic wool industry, England imposed an export tax on raw wool, essentially forcing that wool to be used in domestic manufacturing instead of letting it go abroad. That way the English could export finished clothing products instead of sending out raw materials. England's wool industry remained the dominant core of its textile manufacturing until the rise of cotton displaced it in the early nineteenth century. Wool was so important to England that many of Parliament's import bans in the eighteenth century were specifically designed to protect the industry. Indeed, the rise of the wool industry was fundamental to England's growing prestige in European trade. As Robert Allen puts it:

> By the end of the seventeenth century, about 40 per cent of England's woolen cloth production was exported, and woolen fabrics amounted to 69 per cent of the country's exports of domestic manufactures....Wool was even more important for London. The new draperies flowed out of the capital: cloths amounted to 74 per cent of London's exports and re-exports in the 1660s...and made a large contribution to the growth of that city. By the early eighteenth century, one-quarter of London's workforce was employed in shipping, port services, or related activities.[22]

As London grew, it needed more wood for construction and heating fuel. The overextraction of local forests led to timber scarcities in many local and regional markets, thereby contributing to a sharp rise in timber prices during the seventeenth century. Scholars now generally believe that there was no national deforestation crisis the way economic historian John Nef once argued, but it's indisputable that smaller and more regional levels of deforestation did take place.[23] And these local deforestations had a major effect on timber prices just the same. As a result, the timber scarcities induced a modal adaptation toward the use of coal, which was much cheaper, for industrial activities like smelting and heating.

The annual output of coal in Britain is estimated to have gone from roughly 210,000 tons in the 1550s to almost 3 million tons by the 1680s.[24] Coal had been used by the English for centuries, and it was even known to the ancients as a potential source of heating fuel. But it was extremely dirty when burned, so if timber was plentiful, and it was for most of human history, then people preferred to burn wood instead of coal. Once the English really got going with coal in the late sixteenth century, they encountered a bewildering array of technical challenges that they had to resolve. Households started building bigger chimneys to make it easier for the black smoke to pass through. And although chimneys had been technically around since the Middle Ages, they only became popular starting in the sixteenth century when English homes turned increasingly to coal as a heating source. In smelting, industrial producers refined the structure of the reverberatory furnace to prevent thick coal smoke from polluting the underlying metal. But the biggest technical challenges that the English had to overcome were related to coal mining itself.

The very process of finding and digging through coal mines, especially those that lay deep underground, encouraged the development of one of the most revolutionary technologies in history: the steam engine. Workers in underground mines frequently had to deal with seeping rainwater and uncertain groundwater reserves, all of which could flood the mine and disrupt the extraction process. In 1712, Thomas Newcomen invented perhaps the first practical steam engine in history to help mining companies pump out more water. Although the ancient Greeks had invented the steam engine almost two thousand years earlier, their prototypes were impractical and were not developed any further. Greek societies did not have the same ecological and economic incentives for technological growth that the English enjoyed in the early modern period. Newcomen created a low-pressure atmospheric engine that used the heat from burning coal to boil water inside a large container. The resulting hot steam then pushed a piston that was attached to a mechanical device, often a rotating horizontal beam, which performed a particular task like lifting weights when it moved back and forth. As the hot steam expanded and pushed the piston up, a small jet of cold water was sprayed into the main cylinder through an adjacent valve. That burst of cold water had the effect of lowering the air temperature inside the cylinder. Warmer air holds more water vapor, which is the gaseous phase of water, than cooler air. As the temperature dropped inside the cylinder, the steam started condensing into liquid water droplets. A liquid that forms out of condensed gas occupies a smaller volume and exerts less pressure on its surroundings than the gas. In effect, a partial vacuum emerged inside the cylinder, which means that the pressure inside was lower than the atmospheric pressure outside. The larger air pressure outside forced the piston down, and the cycle continued with a new blast of hot steam. Newcomen's engine was an instant hit because it was far safer and more reliable than previous models.

Some decades later, the engineer John Smeaton made further enhancements to the Newcomen engine and improved its efficiency to a point where more coal mines were clamoring to use it. At roughly the same time, James Watt made other critical improvements. While fiddling around with the Newcomen engine, Watt added a separate condensation chamber so that the main cylinder could remain permanently warm, instead of going back and forth between hot and cold. Whereas the Newcomen engine condensed the steam inside the main cylinder, the Watt engine delivered it to another area through a controlled valve. This simple change more than doubled the efficiency of Watt's steam engine compared to its predecessor. The engine could yield the same mechanical output as the Newcomen models by burning half as much coal, at least. Later improvements substantially expanded the engine's mechanical output. In particular, the development of a "sun-and-planet" gear configuration allowed Watt's engine to convert linear reciprocating motion into rotational motion, making it capable of being used in factories and other industrial places. Incidentally, Watt's new rotational gear system was inferior to an already existing

FIGURE 9.1: Number of Operational English Steam Engines in 1800 by Design Type

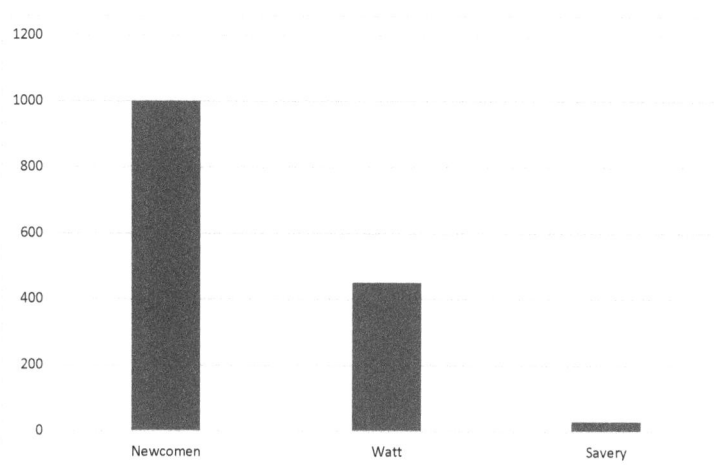

Note: The Savery engines in Figure 1 are older steam engine designs from the English inventor Thomas Savery. These steam engines never really caught on with industry, although a few were produced and adopted by various users.
Source: Lee T. Wyatt III, *The Industrial Revolution* (Westport, CT: Greenwood Press, 2008), 199. See also Sally Baggott, *Matthew Boulton: Enterprising Industrialist of the Enlightenment* (New York: Routledge, 2016), 10.

crank device, but a patent on the latter blocked its potential inclusion in Watt's engines. Once that patent expired in the 1790s, Watt could finally marry the efficiency of his engine with the superior performance of the crank mechanism. Although the sun-and-planet gear allowed for rotary motion, it produced highly irregular and variable speeds that often made them difficult to operate successfully. Watt solved this problem by introducing the double-acting engine, which was probably his most important invention. In a double-acting engine, steam is supplied on both sides of the piston at different time intervals, which produces uniform speeds. Despite all this progress, the aggregate efficiency of English steam engines changed at a glacial pace because Watt and his business partner, Matthew Boulton, secured and aggressively defended a government-issued patent that allowed them to charge exorbitant prices for their comparatively efficient engine. Naturally, many coal companies opted to purchase the cheaper Newcomen engine, even though it was far less efficient. From 1775 to 1800, roughly 450 to 500 Boulton and Watt engines were produced, in contrast to the more than 1,000 Newcomen engines made during the same period. Figure 9.1 shows the composition of English steam engines by design type in the year 1800.

Watt might be remembered as a great inventor today, but he spent much of his career in court proceedings defending his patents from other competitors. Many historians have blamed the patents for slowing technological progress on the steam engine at the end of the eighteenth century, but that assessment largely misses the broader picture.[25] Although Watt's engine was a technological marvel, it still operated through low-pressure steam, which severely restricted the mechanical energy it could generate. The first engines that Watt produced generated only 6 HP.[26] In less than twenty years, his biggest engines were delivering as much as 190 HP, an extraordinary increase in mechanical power, but still less than what was necessary for huge factories and large trains. Watt spent much of his later life waging a very public campaign against the adoption of high-pressure steam engines. The real reason for his vehement opposition is that he viewed them as a threat to his business empire, but publicly he kept focusing on the dangers associated with them, especially their propensity to explode—a valid complaint. Previous attempts at building high-pressure engines had ended in failure, mostly because the energetic steam found its way through cracks and crevices in the boiler, which often led to major explosions. Most British inventors avoided trying to develop high-pressure steam engines for over a century because of these concerns about safety and reliability, even though the technical knowledge for generating high-pressure steam was understood by many.

Watt could not legally prevent anyone from designing and testing a high-pressure engine, and the Cornish engineer Richard Trevithick eventually managed to do just that, ironically as a way of bypassing the limitations imposed by Watt's patents. Trevithick was the real "hero" of the Industrial Revolution, not James Watt, who fought against the technology that would form the basis of virtually all future engines and turbines in human history. Trevithick developed high-pressure stationary engines used by mining companies to extract water and coal out of the ground and then started veering more into steam-powered locomotion. Trevithick's engines dispensed with Watt's condenser and vented steam directly into the atmosphere, generating a characteristic sound that prompted people to call them "puffers." In a major advance, he added his new light engines to carriages, thereby creating the first self-propelled passenger-carrying vehicles in history.[27] He placed the boiler at the back of the carriage and used connecting rods between the piston and the wheels to generate rotational motion. Rail tracks in his day, however, were often not strong enough to handle the weight of these new steam-powered devices, so they broke and cracked quite frequently. Reliable rail would have to wait until the 1820s, when George Stephenson and others built powerful and standardized tracks that could handle steam locomotion. Nevertheless, the derivatives of Trevithick's engine designs gradually found their way into the trains and steamboats that powered the nineteenth century. High-pressure engines were far more efficient than their low-pressure counterparts, but their vast mechanical power is what sealed the deal. No longer reliant on the docile atmosphere to provide some extra pressure, the new engines exploited the dense and energetic steam to push around the pistons with enormous levels of force. Their power, not their efficiency, is what allowed them to move heavy things like trains and ships.

High-pressure steam engines had the advantage of being smaller, more compact, and more thermally efficient. Because pressurized steam contains a higher volumetric energy density than its low-pressure counterpart, every intake stroke of the engine could deliver more mechanical work to the piston for the same amount of steam. They did turn out to be dangerous, as Watt predicted, but far less dangerous than if they had proliferated a century before, when many critical advances in metallurgy were not yet ready. Following

Trevithick and the Cornish engines, Arthur Woolf made the next big impact with the high-pressure compound engine, which used and recovered energy from a working fluid through several stages. In Woolf's compound engine, a high-pressure cylinder vented steam to a low-pressure cylinder, the idea being to give the steam more time to perform useful mechanical work before dumping it to a condenser or directly into the atmosphere. Woolf's engines had far better efficiency than anything made by Watt, but they were often difficult to operate because of their complexity. It wasn't until the engineer William McNaught developed a more practical type of compound engine in the 1840s that it finally became economical for many companies to use steam. From then on, steam power infiltrated virtually all aspects of the British economy.

The diffusion of steam power across Britain was slow at first. In 1800, after a century of development, steam power could produce roughly 35,000 HP, compared to water at about 120,000 HP and wind about 15,000 HP. By 1830, the installed capacity of steam engines was about 160,000 HP, compared to 160,000 HP for water sources and 20,000 HP for wind sources.[28] But by 1870, just forty years later, steam power capacity had reached an astonishing 2,060,000 HP while water capacity stood at 230,000 HP and wind capacity amounted to about 10,000 HP. In other words, installed steam capacity grew at an annual rate of 12 percent from 1800 to 1830 before accelerating to an annual growth rate of 30 percent from 1830 to 1870.

And this brings us to a fundamental point about the early waves of mechanization and spectralization in the Industrial Revolution. Recall that in chapter 5 *exergy* was defined as a thermodynamic system's maximum capacity for useful work. Boosting exergy through the spectralization of conversional technologies was the main causal vector for the corresponding improvement in the efficiency of industrial devices. The Industrial Revolution in England, and virtually everywhere else as well, followed a path of exergy-driven efficiency gains that spurred additional gains and butterfly effects in economic productivity. The English achieved this incredible growth through a huge increase in the aggregate output of mechanical work, a process spearheaded by the spectralization of high-pressure steam engines, and eventually the spectralization of other engine types as well. Let's highlight how this historical process worked through a simple example. In a given unit of time, let's say a steam engine produces 1,000 joules of useful work and has a maximum capacity of 5,000 joules, for an exergy efficiency of 20 percent. However, a more powerful steam engine might generate 10,000 joules of useful work in the same amount of time with a maximum capacity 40,000 joules, for an exergy efficiency of 25 percent. Notice that the exergy efficiency has increased, but so has the amount of useful work produced by the second engine. In other words, *it was not the push for higher efficiency that led to the energetic and biophysical expansion of the British economy, but rather the push to channel and concentrate more exergy, which led to higher efficiencies.* The fundamental energy and economic story of the Industrial Revolution is not about efficiency-driven productivity gains, but about *exergy-driven* productivity gains. The corresponding efficiency increases were purely incidental to the larger project of simply scaling up the productive capacities of the industrial system, which required expanding its maximum capacity for producing useful work. In this regard, Boulton's famous utterance to one of his visitors nicely sums up the core developments of the age: "I sell here, Sir, what all the world desires to have—POWER."[29]

This exergy-driven expansion of the British economy had a curious effect on aggregate-level thermodynamic efficiencies. As steam engines quickly became the prime movers and dominated mechanical output, their gradual inclusion in the economy dragged down the efficiency of the economy as a whole. That's because they started off as extremely inefficient devices compared to more established technologies like watermills and windmills, which had been refined and optimized over many centuries. At the sectoral level, for localized groups of mines and factories, the efficiency of steam engines began at a fairly low baseline with the Newcomen lineage and then gradually increased as future generations of engineers made improvements. But the opposite effect unfolded at the aggregate level. In Figure 9.2, I estimate this era of divergence in the early history of capitalism by specifically tracking the evolution of aggregate efficiencies for English steam engines across time, and I compare that quantity to the aggregate efficiency of English steam engines and watermills combined. The most notable feature in the plot is a sharp drop in the aggregate efficiency of the English economy after 1770, caused chiefly by the Watt and Boulton engines coming to the market and the subsequent expansion of exergy capacity as next-generation steam engines started to proliferate. Although watermills were far more efficient at converting energy than early steam engines, and although watermills vastly outnumbered steam engines throughout the eighteenth century, a typical watermill had a far lower power output than a typical steam engine. That imbalance resulted in a large decline in aggregate efficiency as the number of steam engines increased across the eighteenth century. Note that Figure 9.2 actually shows the aggregate efficiency of steam engines *by themselves* as increasing over time, again with a much faster rate of improvement after 1770 as the Watt engines and Smeaton-enhanced Newcomen engines spread across the English economy. But because these engines were less efficient than watermills and because they start to represent a greater and greater share of all mechanical output, they actually start bringing *down* the overall efficiency associated with both devices. They likely brought down the aggregate efficiency of the entire English economy as well, but the effects there are probably far more subdued as heavy draft animals and other contingents would profoundly affect the results of the aggregation.

This narrative has intentionally focused on steam engines because they were *by far* the most dominant invention of the Industrial Revolution, and their tentacles eventually extended everywhere. By 1835, steam power provided roughly three-quarters of the power output deployed in the cotton textile industry as a whole, or 30,000 HP out of 40,000 HP, with the rest coming

FIGURE 9.2: Aggregate Thermodynamic Efficiency of Eighteenth-Century Steam Engines and Watermills in Britain.

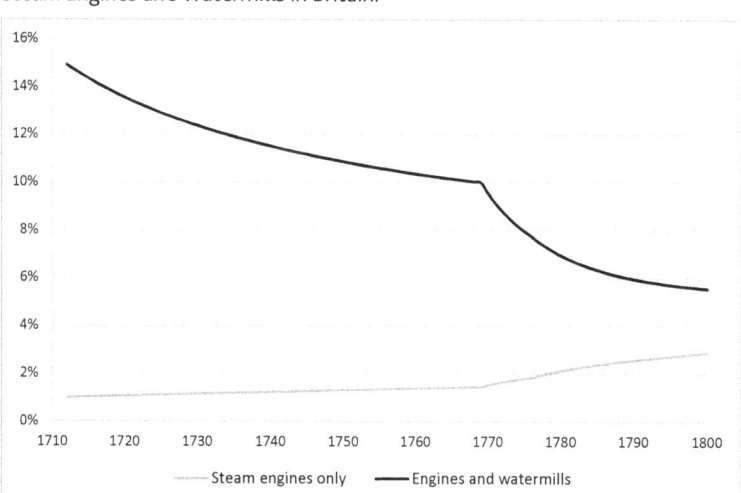

Note: The methodology behind the Figure 9.2 chart makes a number of simplifying assumptions that will affect the final result. First, it does not include the mechanical energy supplied by draft animals, windmills, and other prime movers. A more comprehensive accounting of all these factors would show that aggregate efficiency still declines, but much less than what's shown in the chart above. Second, the methodology assumes that all steam engines and watermills were operating for the same amount of time in every year, which is obviously not true. This assumption will affect the estimated aggregate efficiencies shown in the figure, but it will not affect the basic conclusion that the aggregate efficiency of the English economy slightly declined as energy-intensive but thermally inefficient steam engines came into operation in greater and greater numbers. Third, the number of active steam engines is not reliably known for every single year, so I start off with years where the total number is better known and then use linear interpolation to estimate the active steam engines for the years in between. I also assume a constant 15,000 watermills are operating in every single year. Fourth, the methodology ignores the heterogeneity of the energy qualities involved, as in the case of steam engines coal is burned to generate the heat that produces mechanical work whereas in the case of watermills it's flowing water that's pushing the waterwheel. The annual efficiency of watermills is held at a constant 15 percent for every single year. By contrast, the annual efficiency of the steam engines is assumed to change through some average annual rate interpolated from known values. [30]

from water power.[31] That share kept increasing throughout the nineteenth century. It's certainly true that inventions like the flying shuttle, spinning jenny, and the spinning mule substantially boosted textile production in the eighteenth century, but as with basically every other process in the history of English industrialization, the truly eye-popping productivity gains came in the nineteenth century, and steam engines were the main drivers of those gains. The story of the Industrial Revolution is the story of the spectralization of steam power across different economic sectors. Indeed, one can easily argue that the story of all modern civilization is largely a story about steam power. There's a sense in the public imagination that the age of steam is over and that other technologies have since taken over. The reality is quite different. The dominant prime movers today are powerful steam turbines in power plants that can generate multiple gigawatts as they turn giant magnets and produce huge quantities of electricity. Hot and pressurized steam is still the king of modern civilization, but if we choose to spectralize in a different direction, we may someday succeed in bringing down the curtain on the steam age and raising the curtain on another age of technological progress.

In sum, technological innovation during the industrialization phase of British history was a process of intense conversional spectralization: the interdependent diversification and proliferation of conversional technologies designed to overcome engineering challenges and energetic constraints. The process was interdependent in the sense of being recursive, with outputs from one technological process serving as inputs for others, and in the sense of being socially engineered, with governments, investors, and inventors specifically looking to improve the performance of existing technologies by creating new variations. What typically transpired before modern times is that societies tried to overcome energetic constraints on economic growth by concentrating more human labor, more animals, more machines, or some combination of all these things. In other words, societies tried to overcome energetic constraints largely through modal adaptations. But during the Industrial Revolution in England, investors and inventors came to see the limits and constraints of technological devices themselves as barriers to be overcome and swept aside. They were persuaded to go down this path by virtue of the labor and resource scarcities in the early modern English economy, and by large commercial markets that offered the prospect of huge profits. Productivity gains in the early phases of capitalism relied on the expansion and diversification of mechanical energy. In other words, new technological systems were continuously refined to generate not just higher levels of mechanical work, but also new methods of applying that mechanical energy toward the achievement of different tasks. Watt himself considered the increase in efficiency to be the most important technological innovation of his engine, but efficiency is largely irrelevant during times when resources are plentiful, like coal in Britain during the eighteenth century. The real significance of the steam engine is that it offered people the ability to unleash vast amounts of energy in entirely new ways. Steam power could be used for both transportation and factory work, amplifying the productive and logistical cycles of early capitalism to new heights. After all, capitalists can only sell and distribute a huge number of commodities if they can produce them in bulk and move them where they need to go. The productive power of steam engines, along with many other machines and inventions, offered the new investor classes an incredible opportunity to financially capitalize on the resulting commodity boom. But to do that, the system also needed a vast infusion of money and credit.

War, Imperialism, and Financial Accumulation

Historians like Paul Bairoch have argued that the Industrial Revolution had very little to do with imperialism, colonialism, and global trade. Bairoch made these arguments on the basis that income from activities like the slave trade, and colonial trade more broadly, constituted only a small share of British national income in the eighteenth century.[32] According to Bairoch, Europe also had all the necessary resources for industrialization, so global trade could not have helped much anyway. Bairoch's thesis is flatly false, but not for the traditional reasons that have been given by left-leaning historians like Eric Williams and Joseph Inikori.[33] It's true that Europeans obtained resources like gold, silver, sugar, coffee, tobacco, and cotton from the colonies, but the early waves of industrialization were largely driven by coal and iron, and these two critical substances were sourced within Europe. Furthermore, Spain and Portugal established the first major colonial empires in the world, but they were not the first European countries to industrialize, not even close. Nevertheless, it's impossible to overstate the significance of warfare, imperialism, and strategic competition in the emergence of industrialization. War impacted industrialization through multiple causal vectors. In general, it served as a dominant order parameter that shifted patterns of investment and technological development. The Napoleonic Wars, for example, caused a massive shortage of horses and fodder in Britain by 1811, as the British government prioritized equipping the army with an adequate cavalry force to fight against the French in Spain as part of the Peninsular War. The shortages induced a huge increase in transportation costs, which persuaded many collieries to invest in locomotion using high-pressure steam engines.[34] The first practical steam locomotive entered service in June 1812, and further advances quickly followed in engine technology and railway tracks. By 1830, the Liverpool-Manchester railway line had opened for public use, ushering in a new age in the history of human transportation. And unlike the initial waves of industrialization in the eighteenth century, the Second Industrial Revolution in the late nineteenth century was indeed critically dependent on the violent conquest and extraction of natural resources through imperialism and colonialism, especially since many of these natural resources extended beyond just bullion and cash crops. By the early twentieth century, natural resources like diamonds, oil, copper, cotton, rubber, nickel, manganese, and other minerals were largely extracted outside of Europe and played a dominant role in many major European industries. In 1854, Britain imported 19 percent of its foodstuffs and 26 percent of its raw materials from its external colonies.[35] By 1934, Britain was extracting 42 percent of its foodstuffs and 31 percent of its raw materials from the colonies. In particular, Britain stole and grabbed vast amounts of wealth and resources from India, which was deliberately deindustrialized and impoverished as a result of this systematic looting.[36]

War also played a role in accelerating energy transitions. From 1649 to 1660, in an aggressive naval expansion program pushed forward by Oliver Cromwell, the English navy added an astonishing 103 new warships to its ranks, with 69 of them carrying between 26 to 100 guns.[37] Between 1660 and 1688, the English produced another 106 warships with a total weight of roughly 74,000 tons.[38] By 1688, the navy had become "the most comprehensive and in some respects the largest industry in the country."[39] This rapid naval expansion required huge amounts of timber for ship hulls and iron for things like anchors and nails. The subsequent rise in charcoal prices gradually led to a greater push toward coal-based coke as the main smelting agent in the metallurgical industry. In the popular imagination, Abraham Darby was the first industrialist to smelt iron by burning coke at Coalbrookdale in 1709. In reality, Darby borrowed his methods from earlier pioneers. In the 1690s, Shadrach Fox had been smelting iron with coke at the same blast furnace in Coalbrookdale that Darby would later come to operate. Fox used his coke-smelting techniques to provide cast-iron shells for English forces, which were busy fighting the French in the Nine Years' War. The huge demand for iron and other products for the navy also forced the English to turn to foreign suppliers, in particular Sweden, Russia, and the Baltic regions. At the same time, demand for iron spurred an import substitution strategy in which the English made steady and gradual improvements to iron production techniques, culminating in Henry Cort's famous "puddling process," which converted molten pig iron into wrought iron inside a reverberatory furnace using coal-based coke as the heating source. In 1720, Britain may have produced roughly 25,000 tons of pig iron, a figure that by 1804 had increased tenfold to 250,000 tons, before shooting through the stratosphere in the nineteenth century to reach an astounding 2.7 million tons of pig iron by 1852.[40] Here too Britain's involvement in the French Revolutionary Wars and the Napoleonic Wars proved decisive. Britain manufactured over three million small guns from 1793 to 1815.[41] Pig iron output increased from around 70,000 tons in 1788 to almost 400,000 tons in 1815.[42] The British military "consumed nearly 20 percent of iron output" during this period of wartime, and the iron industry actually entered a period of recession for years following the conclusion of the Napoleonic Wars in 1815.[43] For iron production, textile manufacturing, and many other industries, Britain implemented a policy of import substitution followed by aggressive export-oriented growth. In other words, it first mastered the production of goods that it was originally acquiring from abroad, then it became very efficient at making these things, then flooded continental and global markets with its industrial arsenal through brutal wars and aggressive trade policies. As part of this process, Britain deliberately destroyed and sabotaged India's textile industry to boost its own domestic manufacturers.

From 1692 to 1815, British national debt went from roughly 5 percent of GDP to more than 200 percent of GDP, and this astonishing rise in government debt was almost entirely driven by war and imperialism.[44] Higher levels of government debt impacted industrialization via multiple causal pathways. First, many investors made huge profits in the bond markets of the eighteenth century and the early nineteenth century, accumulating a vast pool of money that was available for investment in the new industries of the era.[45] The most important debt instruments issued by the British government in the eighteenth century were the famous *consols*, short for "consolidated annuities." Introduced in 1751, the consols had a fixed interest rate of 3.5 percent and no scheduled maturity date. Creditors could hold the consols until they decided to sell them on the bond market or until the government decided to pay them off. Issuing consols became a fundamental strategy used by the British government to finance wars in the eighteenth century. Britain redeemed all outstanding consols in 2015. The huge profits from selling consols incentivized the critical investments that accelerated industrialization in Britain, especially in the immediate aftermath of the Napoleonic Wars, when many investors made historic windfalls. Investors would rush to buy cheap debt during wartime, then sold their bonds once prices dramatically increased after the war. The resulting profits were widely invested for industrial manufacturing, spurring productivity and technological growth. During times of war, the British government sold consols at a heavily discounted price from their par value, which is only their face value. For example, if the government wanted to sell a £1000 bond, it would accept something like £500 or £600 for it. As a result, investors rushed like crazy to buy these cheap and perpetual bonds that offered a solid rate of return. After the war ended, consol prices in the bond markets rose sharply because the government no longer issued the cheaper versions of the bonds. The war bonds wouldn't necessarily sell at par value right away, but their price would climb much higher than what the investors had paid for them during the war. Many investors therefore made huge fortunes in the bond markets after they sold these high-value consols, and large portions of these profits were invested in new industries and technologies. A secondary yet important effect of rising government debt was that the nobility started investing heavily in government bonds, leaving less capital for their landed estates.[46] These shifting investment patterns depressed wages in agriculture, prompting peasants and farm laborers to find work in cities instead, becoming part of the new urban proletariat along the way. The capitalists and entrepreneurs active in the new industries of the Industrial Revolution strongly profited from this new wave of cheap urban labor, earning higher profits than what they could obtain in the bond markets in many cases. The financialization resulting from expanded government debt provided downward pressure on wages across the economy and allowed the new capitalists to score higher profits from a vast pool of cheap labor.

Finally, war and imperialism were essential methods of seizing and controlling necessary raw materials for industrial manufacturing that could not be found in Europe. The established formalities of daily capitalism unfold through ostensibly peaceful institutions like markets, contracts, and courts, but the system as a whole is foundationally built upon the violent acquisition of resources by the powerful from the powerless, a historical process that Marx called *primary accumulation*.[47] One could pick dozens of major historical examples to highlight his point, but let's focus on the evolution of the spice trade. In the global trade system of the Late Middle Ages, spices arriving from Asia often fetched huge prices in Europe, where people used them for food preparation and preservation, medical care, and other personal purposes. Before the rise of sugar, wealthy Europeans clamored for vast quantities of cinnamon, nutmeg, cloves, and peppers. For much of the Middle Ages, the spice trade was controlled by Venetians and Arabs, whose legions of wealthy merchants scored large profits. Portuguese naval expeditions starting in the fifteenth century changed these commercial dynamics. Vasco da Gama rounded the Cape of Good Hope in 1498 and entered the Indian Ocean, establishing a major new trade route. The Spanish also joined this commercial arms race. Through military force and economic pressure, Portugal and Spain came to dominate the shipping lanes between Europe and Southeast Asia. Their trade monopolies allowed them to jack up prices on the European markets for critical spices like nutmeg. But the good times did not last. Other European countries became especially jealous of Portuguese profits, and pretty soon English and Dutch ships were sailing all over the Indian Ocean, ready to muscle out the competition. The Dutch East India Company, better known by its Dutch acronym of VOC, established in 1602 as a joint-stock company with the backing of 1,000 wealthy investors, could always count on the full support of the Dutch state, which waged war and sabotaged other European enemies as necessary in support of its crown financial jewel. Dutch involvement in Southeast Asia was cautious at first, but gradually the VOC started establishing forts and military bases throughout the islands that now comprise Indonesia.

Nutmeg was a unique product of trees that grew on the Banda

islands (in today's Indonesia), home to the Banda peoples and their ancient culture. Asian traders had kept the true source of nutmeg a secret for centuries, prompting Europeans to launch expensive and speculative journeys in search of the fabled trees. The Banda practiced neutrality and traded with all foreign powers that reached their islands. The Dutch hated that versatility and developed a simple plan: invade the islands, destroy the local population, and monopolize the nutmeg trade so they could set prices at will. The Dutch assault on the Banda islands in 1621 underscored the ruthless methods of the VOC. The Dutch killed thousands of Indigenous inhabitants and ethnically cleansed the islands.[48] Dutch soldiers and foreign mercenaries beheaded civilians and impaled their heads on pikes; others were simply pushed off cliffs in large groups. In 1620, some 15,000 Banda people lived on the islands. By the time the Dutch were finished in the following year, only about 500 were left. To make up for this lost labor power, the Dutch imported slaves to the islands from abroad. These slaves worked and died on brutal nutmeg plantations, the ownership of which made several Dutch merchants superbly rich. The Banda islands became so important for Dutch commerce that, in a 1667 treaty with England, the Dutch gave up the town of New Amsterdam in exchange for keeping the remote archipelago. New Amsterdam went on to become an illustrious city; today people call it New York.

The increasing success of the VOC in the seventeenth century set the stage for the rise of the Amsterdam Stock Exchange, the first modern stock market. By issuing shares and bonds to anyone who could afford them, the VOC became the world's first publicly traded company. The vast profits of the VOC were traded and packaged through stock options, futures contracts, and other exotic financial instruments. Whereas investors now listen impatiently to decisions from governments and central banks, back then people speculated with their money on the fate of far-flung Dutch fleets sailing with precious cargo. To protect the interests of the VOC, the Dutch state was forced into increasingly expensive conflicts against other competitors, including the Spanish, the English, the French, and the Chinese. Some of these wars turned out well and others did not. Rival European powers eventually managed to steal nutmeg seeds from the Banda islands, planting them in warm regions throughout the world, such as India and the Caribbean. Coupled with increased sugar production from Caribbean slave plantations, this development heavily undermined Dutch control over global trade and started reducing the profits of the VOC. By the eighteenth century, Britain and France had become the new masters of the world's shipping lanes. The Netherlands, a small country that had become a world-beater, went into relative decline and never recovered. The state nationalized all assets and properties held by the VOC in 1799, ending the most successful company in the history of early capitalism.

Precisely because it set the tactics and methods for the rest of the capitalist era, the VOC is perhaps the best reminder that the creation of private property and the protection of wealth usually require systematic and organized violence against weaker groups of people and animals. This may be a difficult concept to grasp given the kinds of economic interactions we have in our daily lives. We find a property for sale, get a loan, buy a house. We try on a dress, see if it fits, pull out the credit card. We hear a good song, open the app, push the purchase button. Violence in our modern economies seems absent when we consider these common scenarios. But the point is not that violence accompanies every single economic or financial transaction. Rather, these transactions unfold in the context of economic systems that were themselves the products of coercion and violence. In order to buy houses, we must first build them on some piece of land. At some point in the distant past, someone probably took away that land from someone else through force and coercion, thus earning the exclusive right to use it and calling it private property. The same logic applies to the very first human inhabitants of that land, who had to displace other animals and plants in order to establish functional communities. Their property may have been shared and communal, but it still required some measure of coercion to actually obtain and control. The arrival of the first humans in Australia, for example, caused the extinction of numerous major animal species throughout the continent. At some basic level, almost all economic activity is predicated on the forceful extraction and isolation of natural resources. However, capitalism transformed this ordinary reality into an unparalleled battleground of violence and destruction, with vast trade surpluses plundered from the colonies and commodified through rigged global markets. Property does not suddenly appear from waving a magic wand; it emerges through warfare, economic pressure, ecological interactions, and political struggles. Then the winners establish laws and customs to make it all seem normal, but those laws are themselves products of the class struggles and labor relations that define the broader economic system. Property is always hostage to the changing fortunes of power.

Concluding Thoughts: The Future of Growth

The vast scale of modal adaptations that followed the Black Death presented new problems, challenges, and constraints for further economic development. Internal and external factors that converged at the right moment launched Europe ahead of other regions in the world. After the energy-intensive modes of capitalism started in Europe, they gradually spread to other regions around the world through trade, wars, revolutions, and migrations. By now, every continent on the planet has experienced varying degrees of urbanization and industrialization. North America, Europe, and East Asia have catapulted ahead of the rest of the world in terms of economic and technological development, but hardly anyone would dispute that capitalism is firmly established on a global scale.

As we've seen in prior chapters, some have claimed that the Industrial Revolution launched a period of self-sustaining economic growth that has no rival in human history. This claim is only partly true. Plenty of economies throughout history have experienced waves of growth and decline. It's not like economic growth magically started with the rise of capitalism in the past few centuries, although this ridiculous story is often repeated by many economists and intellectuals as a way of legitimizing and defending the status quo. What was unique about the Industrial Revolution was *the rapid rate of growth* that it unleashed, a growth wave facilitated by a massive ecological disaster that forced a shift in modal patterns, which in turn led to the mastery and concentration of energy-dense fossil fuels and the resulting spectralization in conversional technologies. But rapid growth rates cannot continue forever because our economies are thermodynamic systems with an expiration date. On a global scale, there are already signs that the rapid growth rates unleashed by the Industrial Revolution are starting to subside. Facing a mounting avalanche of problems that it's incapable of resolving, capitalism is becoming disoriented and losing steam. The time has come to consider a genuine alternative that would ensure the stability and prosperity of human civilization for thousands of years to come.

CHAPTER 10

A New Vision

Until the middle of the twentieth century, there was a widespread assumption among scholars and anthropologists that nomadic life was brutal and miserable, characterized by intense violence, the constant threat of starvation, and a complete lack of leisure time for social activities. By the 1960s, researchers began to study nomadic communities in greater detail, and none received more attention than the San groups in southern Africa. The San have lived in and around the environs of the Kalahari Desert for tens of thousands of years. One particular group, the Ju/'hoansi, was studied extensively by the Canadian anthropologist Richard Lee.[1] His research, and that of others, strongly challenged the most fundamental assumptions that modern civilization held about hunters and foragers. The Ju/'hoansi maintained their nomadic lifestyle by working a grand total of fifteen hours a week. Gender relations were stable and relatively egalitarian, as was the distribution of resources. Sharing was a critical component of San society. When people asked others for some physical item, they usually received it on demand. San communities prized the meat frequently obtained by men through hunting, but they derived 60 to 80 percent of the calories in their diet from fruits and vegetables gathered by women. San children spent the vast majority of their time playing and acting silly, not working or hunting. Group decisions were relatively communal and deliberative; authoritarian tendencies were quickly dissipated and people were strongly discouraged from amassing or hoarding wealth.

The San are not a magical force of human perfection that happened to land in the Kalahari. They are a product of their ecological order, just like the rest of us. From the late twentieth century on, anthropologists have studied dozens of other nomadic communities around the world. Even though these communities had different languages and cultures, they learned the same basic thing from them as they did from the San: people can be happy and content in nomadic lifestyles, even though they lack the vast material surpluses of modern civilization. The fundamental source of human joy is healthy relationships with other people, not the endless accumulation of financial wealth. The modern world has given human beings longer life expectancies, better medical care, and higher levels of education, but also huge amounts of work-life stress and an ecological disaster that threatens the very basis of civilization itself. As we try to understand what a post-capitalist future might look like, it's worth remembering what has worked for our ancestors in the past while avoiding what has failed for our societies in the present. None of this means that we need to go back to being nomads, but we certainly cannot continue being capitalists. Our challenge now is to thread the needle and figure out how to preserve what's great about the modern world while tossing aside those parts that have outlived their purpose.

We are often drowned in reminders about all the great things that the modern world has brought about: longer life expectancy, higher literacy rates, improved nutrition and medicine, lower crime rates—one could go on and on. But there have been some unfortunate disasters of capitalist modernity as well, and it's not just about the brutal wars and ruthless exploitation of our fellow human beings. Consider childcare. For the vast majority of human history, childcare was a mere afterthought; every parent had access to it by virtue of the community in which they lived and functioned. People were surrounded by extended kin networks and communal labor pools that made it easy and convenient to supervise children and to integrate them into their wider social environment. In many parts of the modern world, however, childcare has become just another commodity; it's something people pay for and it's often quite expensive. Many American parents still bear scars and nightmares from the early days of the Covid pandemic, forced by historical circumstances to keep their kids at home while trying to work remotely at the same time. If childcare feels insanely stressful in the modern world, that's because it is. How did this happen?

It came about because capitalism needed an obedient labor force, and capitalists decided to command that obedience by regimenting and regulating our sense of time.[2] For most of human history, there was no division between "adult time" and "child time." Kids grew up in the same environments where adults worked, and thus they learned about what adults were doing from an early point in life. But starting in the nineteenth century, workers faced increasingly stringent requirements for when they had to show up to work, and for how long they had to work. This new world led to in effect a "double division" between adult time and child time and between work time and home time. These divisions had major consequences for traditional family life and profoundly changed the developmental trajectories of young children. In middle-class families in the Western world, such as Britain and the United States, parents regard children as incapable of caring for themselves or other children until about age ten. But in many other communities around the world, children begin to take on responsibilities for other children by age five to seven. Among the Kwara'ae in the Pacific, three-year-old kids are already skilled workers and gardeners.[3] Many children grow produce on these garden plots, then they collect their plants and sell them at the local market, gaining an income for their family as early as three or four. They are also caregivers for younger siblings and are quite competent at social interactions with others. For both adults and kids, work is accompanied by singing, joking, playing, and entertaining conversations.

American parents would not trust kids under five years with a knife. But among the Efe people in the Democratic Republic of the Congo, children could routinely use knives safely, even as infants.[4] Fore infants in New Guinea are able to use fire and knives safely by the time they can walk.[5] Aka parents of Central Africa teach ten-month-olds how to throw small spears and how to use small digging sticks with pointed ends.[6] They teach their young to be autonomous from the infant stage. Infants can crawl wherever they want around camp and are allowed to use knives, machetes, digging sticks, and clay pots. Only if an infant is crawling toward a fire or hits another child do adults interfere. For example, kids just eight-months-old could reportedly use six-inch knives to chop the branch frame of the family house. By three or four years of age, children can cook a meal on the fire. By ten, Aka children know enough about their world to be able to live in the forest alone, if necessary.[7]

For another example, take friendship. It's a historical oddity of capitalist modernity that millions of people are lonely and have no meaningful relationships with other human beings.[8] If you faced a severe crisis in your life, do you have a close friend you could turn to for help in an instant? Many modern people don't, and the question would have seemed silly to most of our ancestors, who were always surrounded by friends and family. Liberal and conservative defenders of the status quo are quick to point out something horrible about the past, like the fact that infant mortality rates were atrociously high throughout human history or the fact that poverty was a near-constant plague for the vast majority of people. But while it's true that premodern humans often faced the prospect of loss and early death, it's also true that they had strong social networks to help them deal with the resulting pain and uncertainty. It was emotionally and psychologically much easier for them to handle a painful loss than it would be for most people today. For all the impressive political, economic, scientific, technological, and intellectual gains we've made, there are nevertheless many things that went horribly wrong in the modern world. Childcare and friendship, for all their significance, are not the real issues. The modern world has brought us the prospects of nuclear annihilation, ecological catastrophe, and massive global wars accompanied by widespread dislocation. Fixing these fundamental problems will require not just tinkering around the edges, but the structural and revolutionary reorganization of all human society.

The Four Pillars

The dominant rhetoric in contemporary capitalism says that we should pursue technological change at all costs. There is very little consideration given to the other major possibility for responding to the ecological crisis and other challenges: changing the organization of society itself. That is now the aim for the rest of our journey: to describe what fundamental changes human civilization needs to make if it wants to survive this millennium and beyond in any semi-recognizable form. Let's begin by highlighting the general principles on which to establish a post-capitalist society. I will identify these principles as *stabilization*, *socialization*, *modularization*, and *localization*; these are the four paths to a better future. It's important to note that these ideas are mutually dependent, meaning that implementing one or two without the rest would be an ineffective response to our ecological and other challenges.

In their seminal 2007 work, *A History of World Agriculture*, the scientists Marcel Mazoyer and Laurence Roudart coined the term *ecological valence* to describe the ability of a species to maximize its population density in different environments.[9] Certain organisms, like bacteria, are capable of living in both normal and unforgiving ecosystems, which is a way of saying that they have a high level of valence. But other organisms require much more restricted environments; you would not find any polar bears roaming around the equator, a sure sign that polar bears have low valence. We will borrow this useful term and modify it slightly for our purposes, redefining ecovalence as the ability of organisms to sustain or stabilize biophysical energy flows in response to external disruptions in their surrounding ecozones. In the context of wild animals, valence could be a measure of their adaptability when they interact with human civilization.

For civilization itself, valence represents the central goal: the protection of a stable way of life in the face of chaotic natural instabilities. I have coined the term *valerism* to capture this new ecological perspective. Valerism is a combination of "valence" and "regeneration." In this context, valence basically means stability; it stands for the collection of stable group modes that maintain sustainable economic activities. Regeneration is the idea that social activities should nurture and regenerate the natural world, not exploit it for short-term objectives. Valerism is compatible with certain forms of socialism and other democratic movements focused on establishing a reciprocal relationship between human civilization and the natural world. The central objective of the valerist state is the pursuit of macroeconomic and energetic stability, making the valerist system very different from capitalism, which is heavily invested in the deceptive prospect of infinite growth. In this context, stability means that energetic scales of production and consumption are changing and fluctuating around some predefined equilibrium. The equilibrium itself could be defined by local conditions, reflecting the confluence of social and political factors that dominate in a particular economy.

On both local and global levels, stability under valerism also means two additional things: that we construct societies that operate within critical predefined planetary boundaries for long periods of time and that we adopt ecological footprints consistent with long-term sustainability. The ecologists William Rees and Mathis Wackernagel introduced the ecological footprint concept, defining it as the amount of land needed to sustain present levels of consumption for an individual, city, country, or any other group of people.[10] For all countries, but especially wealthier ones, stabilizing or reducing the ecological footprint is an essential imperative of aggregate ecological management because greenhouse gas emissions and other harmful ecological consequences are fundamentally tied to land-use dynamics. Whether it's because of deforestation, energy-intensive agriculture, or other related activities, limiting the amount of land we need to sustain our civilizations has to be at the forefront of our ecological planning.

Although growth can certainly occur in a valerist system, in the sense of temporary increases in AFR, growth itself would never be the organizing principle of the economy. To overcome the ecological crisis, and to prevent another one from ever happening again on account of human activity, a valerist economy needs to impose limits on aggregate energy use and consumption. These limits could also be paired with constraints on the consumption of materials and the production of commodities. But society also needs to place limits and constraints on the accumulation of financial wealth, as vast sums of money are often a gateway to accessing more energy for the very rich. I'll begin by focusing on the energetic constraints of the first pillar, stabilization, and after that I'll describe the other pillars.

In the discussion that follows, I'm going to cite energy consumption figures on a *daily per capita basis*. With this standard in mind, the current global average rate of consumption is roughly 50,000 kilocalories, meaning 50,000 kilocalories per person every single day. This number disguises widespread variability among the world's economies. The United States, for example, had an average consumption rate of around 190,000 kilocalories in 2021.[11] Ecological scientists have estimated that, if every individual on the planet consumed resources like the typical American, humanity would need at least three planet Earths to live sustainably.[12] Many other Western economies are generally below the American figure, hovering around 150,000 kilocalories. By contrast, a country like India, the world's second-largest in terms of population, had a consumption rate of about 20,000 kilocalories in 2019. For some historical perspective on these numbers, consider that hunters and gatherers after the invention of fire had a consumption rate of about 4,000 kilocalories. The Roman Empire at its height might have reached an average rate of about 10,000 kilocalories.

This discussion underscores that different countries are facing different realities. In chapter 7, I emphasized that efficiency gains and technological innovation are not the best ways of tackling our ecological crisis. Reducing carbon emissions and increasing fuel efficiency are all great, but global warming is not our only ecological problem and efficiency gains are often used to expand the energetic scale of economic systems. Addressing the ecological crisis *holistically* means that we also should focus on controlling energy use and consumption. However, the controls and constraints that we should adopt may vary depending on the country and the wider historical context that brought it to the current moment. Some countries need to reduce consumption drastically; others can still continue consuming at higher rates for a few more years. But in every society, it's a good idea to establish an *upper* limit of around 60,000 kilocalories for the average energy consumption rate. This limit would be actively enforced through the actions of government agencies, backed up by various constitutional and legal decrees; it should only change in the event of an extreme social emergency. Why should societies choose this range of numbers? There are many reasons, but here are a few big ones. First, it's a reasonable and realistic set of maximum values that would help to constrain humanity's ecological footprint. Second, these energy caps would still allow us to conserve the most important achievements of the modern world, such as higher life expectancies, improved levels of education, and certain features of industrialization. And third, perhaps the most important reason, they are numerical caps that are consistent with the per capita energy consumption figures of certain societies and economies that already generate many positive results for their people.

Societies can also choose to set a lower limit, but here the guidelines can be more flexible. If we wish to protect some of the trappings of modern civilization, such as taking a drive or getting on a flight every once in a while, then a rough lower bound could be something like 30,000 kilocalories. The point of establishing a range, instead of a fixed number, is to recognize that societies are complicated and need some measure of flexibility as they interact with the world and respond to its challenges. Some people may be worried that this range would trap us in a cycle of poverty, destitution, and death. Nothing could be further from the truth. Plenty of well-functioning societies are already in this range, or very near to it. For example, in 2021, Italy had an average consumption rate of about 70,000 kilocalories and Spain stood at around 80,000.[13] The life expectancy of a Spanish citizen is 83 years, and the vast majority of Spaniards are not starving in the streets. It's certainly possible to have healthy societies with far lower rates of energy consumption than those that currently prevail in much of the West and some other parts of the world, and that's because the total amount of energy we use is not the only important indicator of social progress. It also matters how society is organized, how wealth is distributed, how much we invest in healthcare and education, and how we protect our natural environments, among many other factors.

In any case, the only realistic way to impose these energetic constraints is to have strong public and collective control over the dominant sectors of the economy. It's important to qualify this claim and remove some possible misconceptions. A valerist system would still permit the existence of private exchange markets. You can still go to the mall and eat at McDonald's; the government will not take those things away from you. But to prevent large corporations from accumulating too much wealth and power, and to prevent them from becoming energy guzzlers that threaten the planet's ecological stability, the state and other collective institutions should be more actively involved in their ownership and administration. In so doing, the valerist state would also put the brakes on the ruthless tendencies of modern capitalism to plunder natural resources and commodify them for large profits in global markets. In summary, the fundamental features of valerism as an economic system are the following: an average energy consumption rate between 30,000 and 60,000 kilocalories, the organization of economic life around the principle of stability instead of growth, collective and democratic control over the extraction and distribution of natural resources, and a tightly regulated exchange market in which private individuals can obtain higher incomes in the absence of profit-driven exploitation.

Building a valerist future will require economic planning and coordination on an unprecedented scale, bringing us to another core pillar: socialization. We can define socialization as the transfer of assets from private ownership to collective ownership. The more familiar term *nationalization* is merely an approximation, and not a very good one. That's because collective ownership could mean national-level state ownership, of course, but it could also mean other kinds of local public ownership, such as those found in collectives and cooperatives.[14] Because the transition to a new system needs to happen on a rapid timeline, in a historical sense, the market mechanisms that dominate much of contemporary capitalism will absolutely fail to deliver any ambitious results. Indeed, they've already been failing to deliver anything ambitious for decades. In order to successfully implement a program of stabilization, we need a strong measure of socialization. The state needs to have more direct control over the investment decisions made in sectors like finance, defense, agriculture, aerospace, energy, education, and healthcare. The first one of these is critical. By having authority over large banks and lenders, the state will control the financial system not just through a central bank, but also more directly by accepting customer deposits and giving out loans, the same functions now performed by private banks. The current financial architecture of capitalism is thoroughly corrupt and overflowing with criminality from the rich. In the United States alone, financial crimes like predatory lending, tax evasion, and money laundering represent at least hundreds of billions of dollars every year, and the real amounts may be much higher.[15] Putting the state in charge of finance would go a long way toward dissipating the

A NEW VISION

rotten stench of private capital. In the following chapter, I'm going to cover more ground on socialization and the political economy of a valerist society, with a particular focus on the role of the state and on how to make the political and economic structures of this new society more democratic in terms of improving participation and enhancing the decision-making process.

The third pillar is modularization, which is about boosting efficiency by using the same processes and components for different objectives. In general, modularization is the process of dividing a product or system into interchangeable or reusable modules and components. Customization reigns in the current age of capitalism and catalytic spectralization, as consumers are usually sold highly specific variants of commodities that demand specific parts to function. It's a wasteful production method. By contrast, consider something like modular manufacturing, which is focused on producing prefabricated components off-site and then assembling them at a particular location. There are lower greenhouse gas emissions associated with the production and assembly of prefab housing units, for example, than with other forms of housing construction.[16] In a valerist economy, modularization would be used as an industrial and manufacturing strategy for constraining energy consumption and for transitioning technological development from catalytic spectralization toward targeted spectralization. In much of the current public discourse, there has been an emphasis on creating a "circular economy," which is a concept that can mean different things to different people and institutions.[17] In its most dangerous form, it can license the orgies of production and consumption under modern capitalism in the remote hope of recycling the resulting waste products. That's certainly how the corporate world sees the concept: as a useful propaganda tool for faking environmental concern to the public. But in a valerist economy, the primary goal shouldn't be to recycle used materials into other materials. The goal should be reusability, or modular functionality: it should be to reuse existing materials and components in different ways, so that they never have to go through an energy-intensive recycling process to begin with.

Nevertheless, there will obviously be a future need to recycle certain existing waste materials into other usable products, so we should try to understand what modularization and the other valerist principles imply about recycling, a critical industrial activity in the modern world. Waste is the ultimate product of capitalism, and there's no more perfect example of that than the recycling industry. An environmental study estimated that only 5 percent of the plastic waste generated by U.S. households in 2021 was recycled; the rest simply ended up as trash.[18] Under capitalism, we have a recycling industry in name only. In an effort to shift the blame for its destructive ecological activities to consumers, the corporate world has been telling people for decades that they should recycle as much as they can. But it's not profitable to recycle the typical waste products that consumers generate, so corporations simply trash the vast majority of them. If part of a pizza box is greasy and dirty, the entire box is thrown out, even though the clean parts could easily be saved and reused later on. If there's film paper on a cardboard box, the entire box is also thrown out, even though the film paper could easily be removed. Recycling companies in wealthy countries keep labor costs down by dumping most materials in landfills or shipping it abroad to foreign countries, where the vast majority is dumped in landfills anyway. To do otherwise would require hiring tens of thousands of additional workers to sort through recycled materials, and that's not good for the bottom line. No one remains a capitalist by losing money. Recycling is thus a prime example of where one valerist pillar, modularization, converges and overlaps with another: socialization. Simply put, the recycling industry should be a collective and state-owned enterprise geared toward doing its job: recycling as much incoming waste as possible.

Finally, let's talk about localization. In a valerist economic order, production would be localized as much as possible, to avoid using too much energy for logistics and transportation. But it's not just about production; trading systems and exchange markets also need to be more localized than they currently are. Modularization is related to how things are manufactured, but localization is about distribution, about where things end up after being made in the first place. In a valerist economy, we need rational social constraints on product volumes that are allowed to circulate across various geographical scales. The current regime of capitalist production is organized through so-called *global value chains*. These chains are globally segmented structures of production and distribution that ultimately benefit corporations from the wealthiest countries.[19] In the early days of capitalism, raw materials would enter a particular factory and the finished product would emerge from that same factory. Things work much differently now. For example, the components that a particular factory assembles into a finished car are usually produced in dozens of factories around the world and then shipped to their final manufacturing destination.[20] This segmented nature of globalized production is fundamentally rooted in the *global labor arbitrage*, or the replacement of high-wage labor in developed economies with low-wage labor in developing economies.[21] By segmenting production in places with cheap labor, capitalists reduce production costs and score higher profits. On the flipside, international trade becomes vastly more energy-intensive, thus increasingly sabotaging the stability of the biosphere.

For a concrete case, consider the American behemoth Apple, one of the world's dominant tech companies. Apple leverages its superior technological knowledge to design complex electronics, such as laptops and smartphones. But it does not handle the manufacturing process. Instead, it transfers production duties to factories in countries with cheap labor markets, like China and, increasingly, other Asian countries. This strategy has delivered large savings for Apple on labor and manufacturing costs, which would have been extremely high almost anywhere in the West. Taking advantage of the financial mobility facilitated by the regime of global capital, Apple has also established several shell corporations in low-tax jurisdictions, effectively letting it avoid billions in taxes to the United States government.[22] Over time, the company managed to build up a tremendous stockpile of money, at one point reportedly stashing over $100 billion in raw cash.[23] Hardly any of this titanic windfall went into boosting salaries for Apple workers. Beyond the financial fallout, the globally distributed nature of this logistical system requires vast amounts of energy to operate and maintain. Multinational corporations are the primary reason why global trade under modern capitalism is extremely energy-intensive and ecologically harmful.[24]

With these fundamental pillars, we have a broad blueprint for how to move forward. Here are some of the concrete policies and projects implied by this future state of civilization, a valerist society: We would have a universal employment guarantee backed by state funding and organizing, with fewer hours of required work per day. We would provide free and universal healthcare, childcare, public education, and college. We would reduce working days in the week from five to four, along with implementing policies and incentives to diminish any corresponding rebound effects. We would pursue a program of targeted socialization in key industries, like finance, defense, oil and gas, meat and dairy, healthcare, and recycling. We would build more public transportation. We would subsidize the expansion of renewable energy, including building more wind and solar farms but also expanding grid capacity to handle the new energy sources. We would make targeted and balanced investments in improving energy efficiency. We would fund ecological restoration focused on rebuilding and reconstituting vulnerable areas harmed by economic activities. These projects would promote tree and forest growth, natural solutions for reversing dead zones, and silvopasture systems that integrate trees and grazing lands for livestock, boosting productivity and enhancing climate change mitigation along the way. We would move away from crop monoculture systems to more diversified forms of farming. We would encourage more urban farming systems to boost food production. Rather than deurbanizing the world, we would bring the rural to the urban. We would stop the energy-intensive construction of "green cities" and work on making existing cities less disruptive to global ecology. We would promote habitat diversity for critical pollinators, like wild bees and butterflies, to strengthen global food supplies. We would curtail production and distribution methods that specifically generate lots of methane. We would shift toward poultry production and vastly reduce or terminate pork and beef production for being too inefficient and energy-intensive. We would impose financial constraints on individual and corporate wealth, to slow the rate of capital accumulation and thereby weaken a major source of demand for energy services. We would invest in research and production methods to accelerate the recycling of electric vehicle batteries. We would allocate funds and resources to developing countries around the world, both to promote their economic development and to help them manage the ecological fallout associated with the bionomic disruption.

Most of our political and business leaders have come to believe that economic growth is like a magical elixir capable of curing all evils. For many people in the modern world, it does not seem like an alternative to economic growth as currently calculated under capitalism is conceivable. But imagining and realizing these important alternatives is the only way to spare human civilization from a looming disaster. Instead of organizing our societies and economies around the principle of growth, we should organize them around the principle of stability and around sustainable human development, which requires the metabolic cohesion of the wider ecosphere. By tightly constraining the levels of production and consumption around some dynamic equilibrium, respecting core planetary boundaries, stabilizing our ecological footprints, and emphasizing qualitative human-social relations, we can avoid the periodic bubbles and crises of capitalism while also prolonging the duration of human civilization. And by distributing more wealth and resources to workers and common people, we can build a fair society largely untroubled by recurring spasms of political and economic instability. The social and the ecological are inseparable, and together they represent the intensifying political battleground of this millennium. Future generations will judge us harshly if we fail to seize this exceptional moment in history. The impending convergence of crises, from the economic to the ecological, demands nothing less from us than to implement a new vision for our social order.

CHAPTER 11

The Ecological State

The ecological state cannot be abstracted from an ecological society. To analyze the dynamics of the state is to analyze the dynamics of society. Imposing energetic constraints is important for building a society that's compatible with the planet's ecosphere, but we must also explore the economic dynamics that would be necessary to protect such constraints. Because valerism will retain some energy-intensive modes of production and consumption, it will still have a need for collective decision-making and industrial management. Societies wither and decay when hierarchies become static and rigid, or unresponsive to the demands and concerns of the people they rule. In trying to maintain a rigid social order, hierarchy often becomes a source of instability because it denies to people the necessary outlets for addressing and solving critical social problems. The inevitable results for any sclerotic social order are protests, rebellions, wars, and revolutions. This, in essence, is the fundamental problem of conservatism: a maniacal obsession with hierarchical order and stability that breeds more disorder and instability for society writ large. We can imagine a different kind of society with more democratic, dynamic, and flexible social structures that don't freeze class relations in place. A valerist democracy would prioritize democratic representation among the lower echelons and ensure pluralistic methods in the decision-making process. These pluralistic methods can be adopted not just for political governance but also for economic governance, including for collectives, cooperatives, and corporations.

A valerist system needs to prioritize stability by imposing various kinds of rational limits on market activity, capital accumulation, and aggregate energy flows. The main point of these limits would be to prevent the concentration of economic and political power in the hands of a few people, and also to prevent dangerous levels of ecological devastation that could threaten the very stability and survival of civilization. The restrictions would operate at multiple levels, encompassing energetic constraints, financial constraints, and other constraints related to economic scale and activity. There would also be limits on the private wealth that individuals can accumulate. The specific numerical targets can be addressed by each particular society, but they must be consistent with the valerist project of stability, not the capitalist drive for growth. Precisely because it's hopeless to suppose that such stability will emerge spontaneously and organically, driven somehow by magic and markets, we need to clarify the reasons why it should be imposed and the necessary objectives that a valerist state needs to achieve for a successful transition away from capitalism.

The Political Economy of Liberalism and Valerism

Liberal economic theory regards state intervention in the economy as a harmful distortion of the market's apparently inevitable path toward long-run general equilibrium, that magical place where the market satisfies all requests for the right price, the fantasyland where aggregate supply equals aggregate demand. The neoclassical synthesis established at the end of the twentieth century maintains that governments can occasionally intervene to fix temporary problems caused by market activity, but that markets will eventually get it right "in the long run," which is a term of art that economists never specifically define. But even when adopting the myopic and idealized assumptions of neoclassical theory, results from the 1970s showed that general equilibrium is neither stable nor unique, as I explained in earlier chapters. An economy that reaches such a state would fall out of it, and the presence of multiple equilibria leaves open the problem of which one we should aim for. And this objection still leaves out several methodological problems that make it virtually impossible to accurately measure aggregate supply and demand, so one can never really know if an economic system has actually reached general equilibrium, even after allowing for its existence.

But there's an even bigger problem with the liberal conception of the state as the impartial guardian of private property rights, the noble referee of the private sector's mistakes. The state and the accumulation process under capitalism are profoundly intertwined. The state does not merely "protect" private property; it can also *actively create it*. In the 1930s, at the height of the Great Depression, the United States government banned companies from purchasing their own stock, judging this kind of activity to be a form of stock manipulation. But in 1982, after the collapse of the New Deal coalition allowed Reagan to obtain power, the government abandoned the lessons of the past and started to allow a restricted version of stock buybacks.[1] The predictable result was that companies started pouring money into buying their own stocks, sending valuations sharply higher with little regard for actual performance or "economic fundamentals." In the 1990s, the Clinton administration issued new rules about CEO salaries that wound up incentivizing companies to pay their executives through lucrative stock packages.[2] Through these and other actions, the state encouraged massive wealth redistribution toward capitalists and away from workers. Once the apologists of capital took over the state, there was little doubt about who would benefit. And I've already mentioned the well-known example of how expanding national debt turbocharged the process of financial accumulation in early modern Europe, also recognized by Marx in Volume 1 of *Capital*. In particular, the explosion of war debt in the eighteenth century helped to unleash the financial floodgates in many European economies.

These examples demonstrate that the state provides critical top-down constraints on economic activity, and thus exerts enormous amounts of influence over the cycles of production and distribution. The concept of a "free market" is largely an abstraction because virtually all governments have a strong impact on the dynamics of market activity. Governments decide what counts and doesn't count as property; they also enforce those property rights. Governments have considerable influence over monetary supply and the availability of credit; central banks have saved the global financial system from collapse at least twice in the past twenty years. Governments define the rules governing market operations. Governments can even create new global markets for domestic companies through warfare and other forms of strategic competition, like sanctions, embargoes, and blockades. Trade and commerce cannot be decoupled from state power. Likewise, the exercise of state power cannot be decoupled from the class dynamics that constrain the distribution of labor and wealth. The state does not act in a vacuum; its actions are shaped or influenced by various kinds of social and class struggles. The state is a thunderous battleground among competing economic classes and social groups. Economics, especially in the modern world, cannot be understood separately from the collective actions of the state.

The coronavirus pandemic provided another powerful and historic example for understanding the state's critical economic role. In 2020 and 2021, the federal government pumped the economy with trillions of dollars in a desperate bid to save private capital from a systemic breakdown.[3] Meanwhile, capitalists did not hesitate to fire millions of workers as a way of salvaging their profits, while eagerly accepting the trillions of dollars that the state injected into corporate balance sheets. This is the second time in the last two decades that American capitalists have relied on intensive interventions from their government in order to avoid total collapse. How did the workers fare through this crisis? That depends on where they lived. In many European countries, governments took several ambitious steps to prevent economic catastrophe, such as deciding to finance most of the wages for their private-sector employees. Although European nations experienced small increases in unemployment as a result of the crisis, their figures pale in comparison to the initially jaw-dropping numbers that emerged out of the United States.[4] The federalized system produced a patchwork of different responses to the pandemic; this incoherent and uncoordinated strategy was partly to blame for the pandemic's rapid proliferation throughout the country and the corresponding high death tolls. On the financial front, the U.S. government provided money to finance unemployment benefits through a massive stimulus bill, but many workers had a hard time accessing the benefits because of how certain states ran the program. Throughout the crisis, the American people received a painful reminder that the distribution of economic resources, including jobs, is largely a product of social policy, not the preordained outcome of impersonal economic laws waltzing their way through history.

Capitalists run to the state when they need money and favors, but otherwise they want the state to leave them alone so they can continue plundering the rest of society. And there's nothing that terrifies the reigning neoliberal orthodoxy more than the specter of socialization. In the last few decades, many Western nations have sold a substantial portion of their public assets, part of a larger political power shift away from organized labor and toward private capital. These changes may have enriched a few corrupt plutocrats and worsened the lives of millions of people, but they have not altered the strategic and structural importance of the state. Western capitalism seems to be on the verge of collapse about once every decade unless the state intervenes to save the system. When liberal and conservative economists criticize any kind of democratic or social control over the economy, including socialization, they are predominantly worried about how these changes might affect things like productivity and efficiency. We've already seen that these are nebulous concepts with no universally accepted definitions. Consider efficiency. Different research studies focus on different aspects of the term. The general emphasis centers around lowering production costs as one possible method of boosting profitability. For many economists, efficiency has more to do with the "optimal" allocation of resources, such that no new allocation can occur without hurting someone else.[5]

Anti-socialization arguments based on the idea of market efficiency have an extensive history. In 1920, the Austrian economist Ludwig von Mises presented an argument against certain forms of socialism that became known as the "calculation problem."[6] He argued that prices act like signals that tell us about supply and demand for labor and resources. A central board of public planners could never know enough about the fine-grained details of the economy, like how many fish this restaurant needs or how many shingles are going on that roof, to send the right signals to various consumers and producers. Supposedly, only decentralized networks where prices are set between individuals and corporations through mutual consent can offer an ideal allocation of resources.

There are many possible refutations to the calculation problem, but the easiest is to point out examples of complex civilizations that efficiently allocated resources *without using prices at all*. Andean civilizations in South America, such as the Tiwanaku and the Inca, developed complex states and empires without the corresponding rise of a large merchant class. The state controlled the distribution of resources, handing out food and equipment as necessary, and people usually paid taxes to the government in the form of labor.[7] Based on anthropological data, these systems thrived for centuries and they appear to have worked efficiently, in the sense that they consistently avoided extreme resource shortages.

Leaving ancient history aside, markets under capitalism have routinely produced oligopolies and monopolies, creating many inefficiencies and externalities along the way. In other words, capitalism has a tendency to centralize economic planning in the hands of a few powerful corporations, which then control the distribution of resources for other individuals and corporations. Contemporary examples would include Amazon and Walmart, both of which establish prices through central planning for millions, or perhaps billions, of different commodities.[8] Mises was wrong to view prices as innocent markers of supply and demand, as impartial signals about the physical state of the economy. As we've seen, prices usually function like symbolic rituals, as quantifiers of social power that are mediated by class struggles and institutional rivalries.[9] Capitalists price their commodities to beat out the profit rates of their competitors, to seize control over new markets against established rivals, and to extract profits from their hardworking labor force. Capitalists are not always interested in efficiency. They're interested in controlling the social distribution and utilization of economic resources. They're interested in augmenting their power by trying to organize society as they wish, and that process includes pressuring governments and workers to accept their demands through a wide array of threats and coercive actions.

On the empirical side of things, global studies on the relative efficiency of socialization compared to privatization have yielded mixed results. A major study of the British privatization wave in the 1980s revealed no systematic evidence that private corporations were more efficient than the public companies they had replaced. The authors concluded that "it is difficult to sustain unequivocally the hypothesis that private ownership is preferable to nationalization on efficiency grounds."[10] On the other hand, a study about the privatization of Indian banks showed that the national banks were less efficient.[11] Other international studies have offered an equally mixed bag of results.[12] But suppose we grant the claim, despite the ambiguity of the evidence base, that the private sector is more efficient at distributing resources than the government. So what? How does this show that higher efficiency is something worth achieving more than other desirable aspects of economic activity, such as job security and macroeconomic stability? It doesn't. In other words, there are positive aspects associated with greater levels of socialization that we as a society could decide are *worth more* than the negative aspects. These positive aspects include a just distribution of wealth and resources, the restoration of ecological stability, and the improvement of public education, healthcare, and childcare, among others. The efficiency argument against socialization is a waste of time, and especially so from the perspective of an ecological system, which needs the state to have some direct control over the levers of production and distribution as a way of modulating the economy's energy flows.

Levels of public ownership vary around the world. In the United States, public control over vital economic activities and social services is more common than many people think. The federal government still owns the Tennessee Valley Authority, which provides electricity for over 10 million customers.[13] Nebraska enforces direct public control over its electric utility companies, which are governed by "public power districts." North Dakota has a state-owned bank with billions of dollars in assets.[14] Worldwide, governments either control or operate numerous major businesses, including airlines, banks, and oil companies. Finland's government owns Finnair, the country's largest airline. Norway's government owns Equinor, one of the largest petroleum companies in the world. Governments are dominant players in the oil sector, as with Saudi Arabia's Aramco, China's Sinopec, and Russia's Gazprom. Aramco was routinely recognized as the most profitable company in the world throughout the early twenty-first century. Measured by total assets, the four largest banks in the world are all Chinese and all owned by the Chinese government.

These examples are meant to emphasize that there is no obvious contradiction between public ownership and economic success. It's certainly true that many state-owned companies in the past have been operated with great negligence and incompetence, but the same is true for many private companies. How many zombie corporations are kept around by venture capitalists on the fringe promise that they might deliver something in the future, even though they're currently in shambles? How many, like Enron and Theranos, temporarily thrived because of blatant fraud and deceitful behavior? Not only can state companies compete and succeed, they can also provide more stability and certainty to millions of lives that would no longer be subjected to the depravities of capital. State companies do not have to survive by obtaining profits because the government can keep financing them through several different tools and methods at its disposal, including taxation, borrowing, and various forms of monetization, such as printing money. They offer the kind of longevity and job security that private corporations simply cannot match.

Thus far this analysis has ignored something important: history and the geopolitical order. The successes and failures of socialization programs cannot be understood separately from the power dynamics of the global economy. From Iran to Guatemala, many nations challenged the capitalist order in the twentieth century by trying to socialize and democratize the ownership of natural resources. But the core bloc of the global system would have none of it. Because American and European companies were in danger of losing their hefty profits from these socialization programs, the Western powers almost always responded by trying to overthrow the local governments, either through coups and outright wars or by imposing sanctions intended to destabilize the defiant country. We simply do not know how scores of socialization programs would have turned out because they were squashed before having a chance to even get off the ground.

The Iranian example is particularly instructive. Before the 1950s, the production and distribution of Iranian oil was controlled by the Anglo-Iranian Oil Company (AIOC), in which the British government had a majority stake. Rising popular anger about the unfair distribution of profits prompted the Iranian government to nationalize the AIOC in 1951.[15] But the move had many unintended consequences. Britain and other Western countries responded with severe sanctions that made it virtually impossible for Iran to export most of its oil. Iran also lost access to its financial reserves held in Western banks. With the economy reeling and internal political divisions intensifying, the government of Mohammad Mosaddegh was overthrown in 1953 through a violent coup orchestrated by the American CIA and the British MI6. Economic socialization failed in Iran not because of some inherent deficiency, but because the Western powers *decided to make it fail* as a way of protecting their control over the global oil trade.

The precarity of socialization was not confined to smaller economies like that of Iran. The Soviet Union also suffered from the Western-led economic order. Although it was never directly attacked through a coup or a violent conflict during the Cold War, it still experienced the harmful economic consequences of being cut off from multiple credit and technology markets dominated by Western currencies and firms around the world. Despite these restrictions, the Soviet Union still made an amazing amount of progress in various scientific and technological fields, such as launching the world's first artificial satellite and building the first nuclear power plant that supplied electricity to a connected grid. In any case, socialization is likely to be more successful if it manages to expand in the core of the global economic system, particularly in the United States. We need to understand how the exercise of state power can be decoupled from the harmful legacy of capitalism and turned into a positive method for enhancing the ecological stability of society.

The Structure of the Valerist State

The nation-state is the defining political entity of the capitalist economic regime. Scholars typically define a nation as any community of people with a shared language, culture, history, or some other unifying feature.[16] By contrast, the state is a centralized authority that governs people within a particular geographical territory. Nationalism is difficult to define, but it can be generally understood as an ideology and political movement that sees the nation and the state as congruent, meaning that the authority and the borders of the social group should be virtually identical to the authority and the borders of the state.[17] Nationalists generally believe that the interests of the nation outweigh any other competing social interests. Conveniently, nationalists also believe that the interests of their own nation supersede and override the interests of other nations. There are two prominent variations of nationalism in modern times: civic nationalism and ethnic nationalism. Civic nationalism is associated with the French Revolution and holds that a nation is any society of individuals willing to work together for a common purpose, regardless of their ethnic or racial background. By contrast, ethnic nationalism maintains that only people from the same ethnic group should constitute a nation; fascism was perhaps the most extreme version of this style of nationalism. The primary practical effect of nationalism is to suppress internal community conflict while redirecting conflict *outward* from a particular society, which has the wider effect of protecting the wealth and power of elites within that society. Like organized religion, nationalism is a social construct that helps to promote group identity and conformity, making people feel like they belong to a unified community.

The rise of nationalism replaced the political authority of the monarch with the political authority of the people, homogenized and unified through the group modes of capitalism. Culture, language, and religion all became the puppets of capital, which molded and glued them as necessary for the construction of the nation-state. Western historians typically date the birth of modern sovereign nations with the Treaty of Westphalia in 1648, when the European monarchies concluded the brutal Thirty Years' War and established the supremacy of their secular authority over the religious divisions engulfing the continent.[18] In truth, no single event can adequately explain the complicated process that produced modern nations. If forced to pick such an event, the French Revolution would be a good place to begin. In September 1792, with the Revolution heating up and external enemies ready to invade from all directions, French soldiers squaring off against the Prussians at the Battle of Valmy started shouting "Vive la nation," as opposed to the typical rallying cry of "Vive le roi."[19] Two days after the French victory at Valmy, the national legislature abolished the monarchy and proclaimed a republic for the first time in French history. The nation was born in blood and war.

The large commodity markets created by capitalism, spanning everything from the cities to the colonies, could only become integrated through the use of standardized languages, currencies, legal contracts, weights and measures, and complicated financial protocols. The creation of a national consciousness in the wake of the French Revolution relied on universalization and standardization: governments began to mandate that people speak the same language, learn the same history, and participate in the same economic modes of production and consumption. Until the late nineteenth century, for example, large parts of the population of France did not actually speak French. Instead, they spoke a variety of other Romance, Celtic, and Germanic languages. Only a concerted campaign of public education by the state led to the universal adoption of the French language among the French people. Nationalism spread through blood, sweat, tears, and classrooms.

For a century after the French Revolution, nationalism was largely a radical ideology that sparked revolutions across Europe. As the hegemonic core of the world system changed and the new bourgeois classes triumphed, nationalism began to adopt a different role: it became a political mechanism for rallying the middle classes to the defense of capitalism. By giving people a sense of belonging to a community of shared cultural values, nationalism obscured the economic forces and interests that produced the cultural values in the first place. Perpetuating these values and beliefs became an effective way of empowering elite groups that had replaced the agrarian aristocracy. Determining the community of the nation, who belongs and who does not, has always been a complicated process subject to change and revision, following the economic interests of the ruling classes and the agitation of popular groups clamoring for additional rights. In the periphery of the world system, nationalism combined with communism during the twentieth century to construct a broad front of resistance against European colonialism and imperialism. After decolonization had run its course and most African and Asian nations gained formal independence, nationalism devolved into a project of authoritarian rule meant to reinforce the new elites while keeping the rest of the population distracted. Pressured by the hegemonic core, the peripheral capitalists decided that they too wanted to join the global racket. Nationalist leaders began to work on dividing the lower classes through racist attacks against ethnic minorities while also challenging foreign corporations that threatened to accumulate too much wealth in domestic markets.

As long as the nation remains such a powerful vector for accelerating the expansion of capitalism, it will continue to thrive from that expansion at every opportunity it receives. The expansion of capital has structured the formation of national identity through the same universal process: a new capitalist class emerges in some particular society, uses nationalist ideology to reinforce its economic claims against other dominant groups, seizes control over the distribution of wealth, then modifies nationalism to reinforce and legitimize the hierarchies of the new system. Nationalism has turned into the useful idiot of capitalism, which will never abandon its servile puppet unless forced to do so through war, struggle, revolution, and a complete transition to another economic system.

What is to be done about the nation in the future? On the one hand, it seems inconceivable to abolish the nation-state; it's become the political equivalent of breathing air. And yet, nationalism cannot be the guiding political philosophy of the valerist order, since it's painfully obvious that some collective interests vastly outweigh the interests of any given nation. If the biosphere collapses, all nations go kaput. When carried to even mild extremes, nationalism also drives geostrategic rivalries among competing nations, which start to see their relationships increasingly through an oppositional and antagonistic lens. The central issue is not over how to abolish the nation, but how to save it from itself, and especially how to resuscitate it from the corrosive social and economic entanglements of late-stage capitalism. A great place to begin is by reconsidering how democracy can reinvigorate the nation in the new valerist framework.[20]

In modern times, elite opinion in the West has come to define democracy as a political system with competitive multi-party elections, checks and balances among different coequal branches of government, and an independent judiciary that administers the law free from political interference.[21] This form of democracy is usually called *liberal* democracy, and it's the monolithic version typically presented in the Western press. But many other forms of democracy exist and have existed in the past. In *Politics*, Aristotle was quite clear that a democracy is a society in which poor people are in charge of the government, not the rich.[22] For him and others in Ancient Greece, societies that used elections to fill the offices of government were actually considered *oligarchies*, because elections are easily susceptible to being corrupted by the rich. One need only look at the modern United States as an ideal example: judicial elections at the state level are often bought and paid for by large corporations, vast sums of money virtually dictate the outcomes of most congressional elections, and it's practically impossible for a regular person to obtain high office without groveling at the feet of some capitalist overlord.[23]

In contrast to oligarchies, many democracies in Ancient Greece used the method of sortition, or random lotteries of their citizens, in order to fill legislative positions in government. It's the same basic process used in the United States to select jury pools. Of course, these societies had their own major problems, including slavery, poverty, imperialism, and so on. But the basic point is that elections are not the only way to fill political positions. Although political systems can certainly be distinguished by the rules and procedures they follow, state typology in Ancient Greece usually depended on the distribution of power in society. And this is the more holistic conception of democracy compatible with valerism as well: a political and *economic* system that distributes power broadly across multiple groups and classes. This is a radically different notion of democracy than what most people in the Western world have become accustomed to. The political philosopher John Burnheim introduced the concept of a *demarchy* to talk about a style of democracy that concentrates decision-making power in the hands of people who are most affected by the issues in question.[24] The implication is that not all major decisions should rest in the hands of politicians who are far removed from the particular problem they're trying to address. The idea is to replace central authority with more local forms of management, networking, and communication. There is hardly a better method of achieving this noble goal than sortition, since sortition places people in power who are directly impacted by political action.

Although local solutions to complex problems are always desirable when they can be found, managing the global ecological commons will inevitably require some kind of central authority. Frequently in their history, societies need to concentrate massive amounts of political power to remove critical pressure points and get something ambitious done. This is one of those times. During the Second World War, the War Production Board in the United States restructured the country's industrial output to focus on producing weapons. Private companies were forced to stop their regular activities and switch instead to making weapons, from rifles to tanks. It was a brilliant success; the United States produced around two-thirds of all the military equipment used by the Allies during the war.[25] For just some examples, the United States produced a total of almost 300,000 aircraft and 86,000 tanks. And as the case of Japan under the Tokugawa highlights, the interplay between local and central dynamics is critical to completing any noteworthy ecological project on a grand scale. Simply put, local networks by themselves are not capable of organizing society for large-scale projects that implicate the survival of the global economy and human civilization.

Our primary goal then is to design and implement flexible and dynamic levers of political control in the nation-state, as a springboard to more democratic forms of governance for the global commons as a whole. There are several ways of achieving this goal without devolving into crude forms of dictatorship and authoritarianism. The first is to focus on the constitutional structure of the valerist state. Nation-states can deploy political power through many constitutional systems, but they do so most efficiently through a unitary system; in such a system, constitutional powers are exclusively concentrated at the national level and are not shared or distributed among multiple layers of government. The vast majority of the world's countries are already unitary states, but not all of them. Countries like France, Italy, Japan, and China all have unitary constitutions. Local governments exist and proliferate in unitary systems, but they derive their power from the national government, not directly from the national constitution. In a unitarian state, the chain of political command is explicitly clear, and it's therefore easier for national governments to reorganize society and the economy when faced with a major challenge.

One of the constitutional hallmarks of the modern liberal state has been the tripartite division of power among three branches of government: the executive, the legislative, and the judicial. This liberal triumvirate is useful because it distributes government functions and strikes a nice balance between the exercise and the limits of power. At the same time, the balance of power has swung too far from the people and the triumvirate needs to be heavily revamped. In particular, the notion of coequal branches takes a great deal of power away from the democratic will of the masses and puts far too much power in the hands of unaccountable or unelected individuals, such as judges and bureaucrats. In a valerist democracy, the legislature must clearly and unambiguously be formalized as the most powerful branch, with the other two being merely junior branches for either executing or interpreting its will. This idea was indeed the intent of many of the founders of the United States, hence why Congress prominently receives top treatment in Article One of the Constitution.[26]

THE ECOLOGICAL STATE   337

To the extent that a constitutional order is necessary for underpinning the political system, the main role of any given constitution should be to explicitly spell out the dominance of the legislature relative to the other branches. The legislature should dominate the political system because the democratic legislature is the voice of the people. In this system, for example, the main purpose of the judiciary would be to interpret the intent of the legislature in ambiguous situations, not to decide whether a particular law is constitutional or not. Only the legislature would have the power to reverse or change laws; if one particular legislative session gets it wrong, another one would have the chance to fix that mistake later on. Crucially, no court would ever have the power to overturn a law; courts would only have the power to decide how laws are applied in specific circumstances where there's no clarity in the text. The kind of judicial review that has poisoned the political system in the United States would be abolished and buried deep underground. In addition, legislatures should be unicameral. In the practical and historical experience of most countries around the world, bicameral legislatures have been either dominated or corrupted by conservative upper houses, like the Senate in the United States or the House of Lords in Britain. When it comes to the major or controversial provisions of any bill, the Senate in the United States pretty much dictates the outcome to the House of Representatives, which has little structural power in the American legislative system because, simply put, it's relatively easy to get things through the House but difficult to get them through the Senate, given its internal rules and constitutional structure. The end result is that any semi-serious bill that emerges from the House is watered down by the more conservative members in the Senate, making it difficult to implement an ambitious political program. To ensure the effective transmission of democratic power throughout the political system, it's therefore absolutely essential that national legislatures be organized as unicameral institutions, with only a single house present in the legislative branch.

A valerist democracy also needs new groups of institutions and social formations for managing the increasingly complex problems human civilization will face in this millennium. At a more local and granular level, I propose the creation of social-political networks of collective management over economic and ecological resources. I will call this political unit a *demopole*, short for *democratic polis*. Demopoles will function as executive councils that contain members from their local districts. Within reasonable constraints related to the removal of dangerous criminals and the imposition of age thresholds, all citizens of a given country would be eligible to serve as demopolists in the demopoles. The demopoles would be assigned territorial and geographical districts of control; their goal is to manage economic and ecological resources in accordance with the authority and strategic direction they receive from their respective national governments. They will, in effect, function like worker's councils with broader political authority over local economic decisions. Demopoles will blend both political and economic roles, and members in each respective demopole shall be chosen through sortition, in line with the more ancient and traditional understanding of democracy. At least in this one respect, the valerist project is radical by aiming to restore something important that's been lost from the past.

Demopoles are meant to be subnational units of governance, and thus they are subordinate to the laws and regulations passed by their respective national legislatures. I'll frequently refer to national legislatures as *mesopoles* for reasons that will become fully clear in chapter 12. Essentially, they're being called that because they'll be acting as a middle layer of governance in the valerist global system. Mesopolists are to be selected through sortition as well, and here is where sortition would automatically eliminate one of the fundamental problems of modern legislatures and constitutions: corrupt districting rules that give minority factions too much power relative to majorities, such as the kind of gerrymandering that's become notorious in the United States and other countries. Because all eligible citizens anywhere in a particular country can serve as mesopolists, the random lotteries would unfold on a national scale and would therefore eliminate the need for dividing political districts. Of course, the division of political districts can and will still happen for many other reasons, but it would no longer be necessary specifically for the reason of selecting mesopolists to their national legislatures. Coupled with valerist mesopoles that are unicameral, we would truly have a political system fundamentally built on democratic representation and the will of the people.

Elections do not have to be totally eliminated in a valerist polity. Perhaps it's still worth preserving some kind of electoral system for selecting executive leaders at the national level, and for selecting executive leaders at other local levels of government as well. It's true that these elections will still be susceptible to being corrupted by the rich, but the fundamental difference now is that the national legislatures, the mesopoles, will have supreme powers over the structure of governance in the valerist state. Even if an executive leader is captured and manipulated by a corrupt faction of wealthy individuals, which is something that will likely happen from time to time, the system will be structurally tilted toward the exercise of power by the mesopole, and so a corrupt leader will not be able to do much damage. Like any other political system, valerist democracies will have their own flaws and imperfections, because human beings are always a work in progress. But at the very least, they'll be much better than our current systems at implementing the democratic will of the people, at effectively channeling political power for the completion of massive projects, and at addressing the global bionomic disruption and the associated ecological challenges it has brought.

What I have outlined here is a rough blueprint for an ecological state, a government that takes a holistic look at social problems and decides to elevate the importance of ecological concerns in addressing those problems. This kind of society promises a far brighter future for humanity than the current system. We can do better, and we should. But to achieve anything meaningful, we must turn our attention to the changes that need to be made to the international order. No nation is an island; countries interact with each other in many complex ways, and we can't achieve ecological success at a global level unless we change the rules and institutions that currently prevail in the global capitalist order. These rules are supercharging violent conflicts, economic competition, technological rivalries, and the emergence of oppositional power blocs in global affairs. To break this cycle of corruption and failure, we have to pursue effective alternatives that are compatible with human civilization's long-run survival in the biosphere.

CHAPTER 12

The Valerist Cosmopole

The world is an integrated community where decisions over here can have a huge impact on what happens over there. In this geopolitical context, replacing capitalism also means replacing the current international order that sustains its existence. It means designing new rules and institutions that foster global cooperation rather than enshrining plutocratic and corporate hierarchies that benefit the few. To manage the global commons in an age of ecological convulsion and to undo the depravities of capitalism, we need to create a new international order that will ensure the long-term survival and success of human civilization.

In 1968, the ecologist Garrett Hardin wrote a famous article about the dangers of overpopulation and the exhaustion of our ecological resources.[1] If people are given free access to a common resource, he argued, they would quickly deplete that resource and there would eventually be nothing left. Herders would exhaust the grass in common pastures by adding more cows, fishers would catch all the fish and deplete existing stocks, people would clear out all the forests to satisfy their need for timber, and so on. Hardin suggested that the solution to the problem was either privatization or government control, in the sense that both options would prevent people from accessing the resource in question.[2] But it turns out that's a false choice. Throughout history, many people in local communities across the world have managed common resources through negotiation and collaboration.[3] As the economist Elinor Ostrom also explained, multiple other paths exist for the communal management of natural resources besides those presented by Hardin.[4] Her research demonstrated that people can successfully manage common resources through collective action at the local level, and she offered several design principles for managing these common-pool resources. For a few examples, the rules governing the use of common resources should be adapted to local conditions, as many resource users as possible should participate in the decision-making process around the resource, the utilization of common resources should be carefully monitored, and sanctions for violators should be gradual and incremental rather than immediately harsh and punitive. Although Ostrom's work offers many important lessons, it's useful to remember that the planetary and ecospheric problems we're currently facing do not always have easy analogies to local issues that can be resolved through communal planning and negotiations.

On a global scale, the socialization of economic resources and the democratization of public life cannot happen in the context of our current international order, which is based on wealthy countries in the West exploiting cheap labor and extracting natural resources from poorer countries in the Global South and other parts of the world. What we need is the creation of new governing institutions for ecological management, conservation, and restoration, meaning we need a new network of institutional arrangements and agreements for managing the planetary ecosphere. We also need international mechanisms for enforcing the dynamical constraints on economic activity that will be imposed under a valerist system. This global effort will involve both conflict and cooperation among different nations, organizations, businesses, and various social and political groups, but the ultimate goal is to transition from the economic order of capitalism to the ecological order of valerism.

The Current Structure of the International Order

The current international order does not bode well for the future of human civilization. Multinational corporations are plundering the planet for critical natural resources and causing vast social and ecological disruptions along the way. Most national governments are happy to look the other way while big capital exploits cheap labor, flattens tropical rainforests, pollutes the world's oceans, and accumulates obscene levels of wealth from the ongoing ecocide. What's worse, the intensification of geostrategic rivalries, especially the rivalry between China and the United States, will make it all the more difficult to have an effective ecological transition. It's not just expensive to fight wars and maintain large militaries; it's also energy-intensive and ecologically destructive. The United States military is one of the worst polluters in the world, pumping enough greenhouse gases into the atmosphere to rival the emissions of most midsize countries.[5] And China, for all its amazing investments in electric cars and renewable technologies, will feel enormous pressure to continue burning huge quantities of fossil fuels as a way of ensuring economic stability in the face of a chaotic geopolitical environment. As of 2023, China is by far the largest emitter of greenhouse gases in the world, although Chinese per capita emissions are still lower than American per capita emissions.[6] The rivalry between the two superpowers is also generating an intense race to control rare earth minerals and other critical resources that are necessary for electric car batteries, smartphones, and the technologies of the future.[7] A Cold War II between the world's two major superpowers would be an absolute disaster; it would simply turbocharge the economic absurdities of our current age along with the rapacious ecological destruction of the planet, thus inevitably hastening the collapse of global civilization. China and the United States should do everything to avoid it, including reaching a grand strategic bargain about how they can divide and distribute power in the context of a multilateral global system.

These ongoing geostrategic rivalries reveal that capitalism still thrives from waves of imperial plunder, agitation, and competition. Various forms of imperialism have existed throughout human history. The world has witnessed the adventures and depravities of the Assyrian Empire, the Persian Empire, the Roman Empire, the Mongol Empire, the Ottoman Empire, the British Empire, and many others. When most people today think about historic forms of imperialism, they tend to think about the domestic political system of the country in question. This is certainly how many political scientists operate; they tend to blindly assume that an empire is defined by its internal dynamics—by whether it has a hereditary monarch or an elected president, whether it has checks and balances in its political system, whether it allows for certain kinds of human and civil rights, to name just a few examples. But an empire is better understood by its external relations because those more accurately determine how it interacts with other nations; in other words, it's better to study an empire by looking at how it acts vis-à-vis other countries. This has been the general approach behind the greatest studies of imperialism, including those conducted by Lenin and the scholar Immanuel Wallerstein.[8]

In this sense, empires are powerful states with satellite clients, puppet regimes, spheres of influence they won't allow others to breach, and extractive economic relations that benefit the imperial core while harming the periphery. Just about any kind of political system can exhibit imperial behavior given the right circumstances, regardless of whether we call that country a democracy, an autocracy, or something else. Some empires, however, will obviously be more powerful than others, and a set of nested hierarchies can arise in international relations, whereby smaller imperial zones will defer to bigger imperial whales on the big questions of the global order. Immanuel Wallerstein, for example, saw the capitalist world system as divided between an imperial core and an externalized periphery, with unequal economic exchanges prevailing between the two spheres.⁹ The global economy is fundamentally based on cheap labor from countries in the Global South, providing raw materials to countries in the Global North, which then converts these raw materials into finished manufactured products that are sold at high prices in global markets. In the chocolate industry, to pick an example, cocoa beans are grown in West Africa, but the chocolate bars and products that consumers buy at the store are manufactured in places like Europe and North America. Much of the Global South is therefore deliberately kept deindustrialized and underdeveloped, for the simple reason that dominant capitalists in the global economy prioritize the cheap extraction of raw materials in order to sell finished products at expensive prices. And this basic dynamic is reinforced and reproduced in various ways through global institutions.

The current global order was largely created in the aftermath of the Second World War and therefore reflects the power dynamics that emerged from the war. After much of Eurasia was destroyed in the war, the United States emerged as the world's primary superpower and went on to establish a series of international institutions to extend its political and economic control over the global system. Organizations like the IMF and the World Bank were officially designed to stabilize and help the global economy after the war, but they've instead become blunt instruments of American power. By the 1980s, the postwar global economy had evolved into the "Washington Consensus," the pro-market capitalist status quo that emphasizes privatization, cuts in public budgets, and free trade, which practically means poor countries opening up their economies to competition from advanced economies.[10] These postwar financial institutions and arrangements are therefore tied to maintaining and supporting a corrupt economic order based on capital accumulation and labor exploitation. In particular, the structural adjustments programs of the IMF have been abysmal failures, imposing widespread fiscal austerity and significant cuts in public services on vulnerable nations that are still trying to lift their economies off the ground. From Mexico and Russia to

Greece and Argentina, the IMF has left behind a streak of disastrous interventions that have induced multiple deep recessions and depressions, leading to sharp reductions in living standards and causing political instability along the way.[11]

The United Nations also emerged from the ashes of the Second World War, meant as a more effective replacement to the League of Nations. Although the UN system does have many advantages over its predecessor, it's not as effective as many people seem to think. Wars and conflicts declined after the Second World War largely because the United States became the leading hegemonic power in Europe, and thus restrained its European satellites from engaging in the kinds of brutal conflicts that they'd been carrying out for centuries. The United Nations does provide a useful forum for deliberations and conversations among the world's countries, but it still has no effective enforcement mechanisms. Beyond that, the central flaw of this system is that just any one of permanent members of the Security Council—France, Britain, Russia, China, and the United States—can sabotage the will of the entire international community. The broader Security Council does have fifteen seats, ten of which rotate between different countries, but the effectiveness of this otherwise admirable rotating system is thwarted by the five permanent members, which are the ones that really decide what does and doesn't happen. Instead of looking like a global assembly that reflects popular will, the United Nations looks more like an exclusive club for big countries and two irrelevant European bit players that once had large empires.

Our goal should be to establish a new international order that's more inclusive and pluralistic. We'll never get to a utopian ideal of eternal peace on Earth; human beings are too complicated and too often driven by various jealousies, rivalries, and other antagonistic factors, even if a well-constructed socioeconomic order can temper and subdue some of these harmful impulses. Conflicts and disagreements of various kinds will always be part of human society. And that's precisely the point: we need new institutional frameworks that can more effectively limit the harmful impacts of wars and conflicts, that can properly manage disagreements among nations, and that can collectively manage the world's ecological resources for the long run. It's this last imperative that should drive our future plans, since we've seen how ecology is really the central basis for the major features of civilization. It's impossible to know how this new order may come about. It could emerge through long-lasting negotiations among responsible national governments. It could emerge through a hybrid approach involving governments, businesses, and civil organizations. It could emerge from revolutionary waves that sweep aside the current order all over the world. Humanity is dynamic and unpredictable; it will figure out a way.

The Cosmopole and the Living Earth

It's painfully clear that our systems of global governance need to radically change if we're going to have any hope of making global civilization compatible with the wider ecosphere. These changes must address two primary deficiencies: first, that the current global system is politically ineffective when it comes to ecological management because it's rooted in nationalism and capitalism; and second, that the system doesn't reflect the democratic aspirations of the masses worldwide, meaning that the people currently making major decisions about the collective future of humanity are not representative of what most of humanity thinks, believes, and desires. To remedy these problems, it's necessary to think big. At the global level, the main political vehicle for promoting the new valerist order will be what I'll call the *Cosmopole*. The Cosmopole is not designed to be a silly abstraction about world peace and harmony. It's meant to be a political assembly and global legislature that's composed of members chosen through sortition from its constituent mesopoles and demopoles. In periodic time intervals, perhaps once every three or four years, a portion of the Cosmopole would be rotated out and a new set of members would replace the outgoing ones. To account for differences in experience, mesopolists would be given the greater weighting in the sortition, meaning that they would comprise a majority of members in any incoming rotation. The Cosmopole would be established through a founding charter approved by the world's nations. Decisions reached by the Cosmopole are meant to be binding on all national governments around the world.

There are many important features that would make the Cosmopole fundamentally different from current structures of global governance, such as the International Monetary Fund or the United Nations Security Council. First and foremost, the allocation of seats to the Cosmopole would be based on population at the national level. For example, India had a population of roughly 1.4 billion as of 2023, when the global population was about 8 billion people. That means India would have received 17.5 percent of all the available seats in the Cosmopole. By contrast, in the current international order, India has neither a permanent seat on the Security Council nor any significant influence and voting power at the IMF. The largest country in the world has virtually no influence over major international institutions, a clear absurdity reflecting the obvious fact that the current international order was based on what happened almost a century ago, in the aftermath of the Second World War. Second, the Cosmopole would function as a legislature that passes binding laws, and these laws would be executed by the world's two hundred national governments. There would be no corresponding executive or judicial functions at the global level, meaning that the will of any given Cosmopole cannot be overturned except by another legislative session of a later Cosmopole. No court system of any kind, national or subnational, could reverse any law passed by the Cosmopole. Third, the Cosmopole would have the power to control, regulate, establish, or fully abolish any international institutions that deal with global trade and finance. For example, it could legislate the IMF out of existence if it so chose, or it could reform it through various measures designed to make it more effective and democratic.

The founding charter would also spell out the constraints

governing the functions of the Cosmopole. For example, the Cosmopole would not have the power to directly set tax rates or conduct monetary and fiscal policies; those would still be reserved for national governments. Nor would the Cosmopole have the power to weigh in on social issues like abortion and same-sex marriage; those powers would also be reserved for national governments. However, the Cosmopole would have expansive powers in areas that indirectly affect fiscal and monetary policy. The Cosmopole would be created, chartered, and authorized primarily for the purpose of global ecological management. It would have the ultimate power to halt, initiate, or approve major construction projects, restoration projects, technological projects, and other economic activities that could significantly affect aggregate energy flows. If a country decides to issue drilling rights to an oil company in sensitive lands, the Cosmopole would have the authority to overturn that decision. It would have the authority to cancel or modify large real estate development projects. The Cosmopole would have the power to establish new public parks that are off limits to economic development, or to modify the boundaries of preexisting parks. The Cosmopole would have the authority to overturn judicial decisions at the national level that are inconsistent with its mission of global ecological management. These are just some examples of what it could do under its general purpose and strategic vision.

The nation will thus have both a new foundation, the demopoles, and a new guiding hand, the Cosmopole. It will be subsumed in a new valerist superstructure for managing the global commons. This approach would conserve the existence of the nation-state, but would remove nationalism as its guiding philosophy. It would recognize that national interests are not diametrically opposed to the collective interests of humanity, but are indeed aligned and compatible on the most important issues facing civilization in this millennium. In this way, the Cosmopole will serve as a dynamic and responsive institution for integrating the world's population in a broader global democratic project.

Certain critics and right-wing demagogues will scream "one-world government" at this plan, but they would be wrong, because this plan does not call for a single executive branch to rule over the entire world. In other words, there would be absolutely no risk of a demagogue or dictator emerging to run the whole planet, because there would simply be no executive position at the global level. The Cosmopole will legislate and the world's two hundred nations will execute. In this valerist order, legislative power is concentrated at a global level but executive power is spread out among all the world's countries. There is no judiciary necessary to check the authority of the Cosmopole, as its membership will be replaced through sortition and modular rotation over time. The dynamism of the Cosmopole is inherently a check on its power, as no Cosmopole can be easily corrupted by outside forces and future Cosmopolists can always overturn the poor decisions of their predecessors.

One of the central challenges with this new international order revolves around enforcement. The League of Nations was ineffective because it had no mechanisms of enforcement; nations invaded other nations and nobody really cared about what the League said. In some sense, the United Nations suffers from the same basic issue. It's not a governing body that can compel countries to act in certain ways. The usual problem with international bodies is that their words are manufactured by lawyers, but armies are controlled by generals. How can a nation be compelled to do something if it just doesn't want to, the international system be damned? To grapple with this question, it's useful to begin by acknowledging that nations are confrontational under capitalism because there's a global race to the bottom. Conflict and confrontation are part and parcel of the capitalist project, which is deeply rooted in the rapacious extraction of natural resources and the exploitation of cheap labor. A different economic and ecological order assumes that the class dynamics of capitalism will be long gone, that political systems will be responsive to the needs of their people instead of their billionaires. War will still occur in the valerist international order; there's no point in having unrealistic expectations about the future. However, war will be much more difficult to carry out for a lengthy period of time, and a global prohibition on the possession of nuclear weapons would also constrain its most harmful effects. The nature of democracy under valerism will ensure that all layers of the global order are fluid and dynamic, responsive to changing economic and ecological conditions.

Several ideas or mechanisms could facilitate the Cosmopole's effective enforcement over the global commons. First, national governments could enshrine the legal jurisdiction and authority of the Cosmopole in national constitutions, meaning that disobeying the Cosmopole would be tantamount to explicitly violating your own constitution, with all the attending political consequences that such a move may bring. Second, the Cosmopole could control the fate and structure of institutions like the IMF or other global financial bodies, meaning that it would have enormous economic leverage over specific countries and more broadly over the global economy. It would be unwise for any nation to directly challenge that level of influence, though certainly some will try. Third, the Cosmopole could decide to punish offenders through changes to its own internal rules of membership and operations. For example, perhaps repeat offenders could receive temporary suspensions of their membership in the Cosmopole, assuming that a majority of the Cosmopolists vote to take that decision. There are lots of possible options here, but the broader point is that, because of its unique institutional structure and central place within the global system, the Cosmopole would have plenty of tools to ensure that it remains effective and relevant.

We now have a plausible international path forward for the future of human civilization. A responsive international order would make the successful evolution of the entire biosphere a political project. Nature would no longer be a stranger, a foreign entity, or some "Other" that has to be defined in opposition to the social world. Instead, the fate of the social world would be indisputably tied to the fate of the natural world as a whole, for only the betterment of the latter can facilitate the success of the former.

There will always be nature without society, but there can never be society without nature. Being mindful of our ecological limits and acting accordingly would be a testament to our civilizational maturity.

The future is unknowable, but not always uncontrollable. Humanity now has a historic opportunity to control its destiny. Although it's true that we are generally better at managing short-term problems rather than long-run challenges, it would be a severe flaw to blame some vague and amorphous sense of human nature, and all of its supposed foibles, for our current predicaments. Human nature is not a static wall impregnable to outside circumstances; it's a dynamic, living and breathing force stretching across the eons, shaped by external and internal conditions of all kinds. We have the power to reshape our political and economic orders, and in the process to reshape our very nature: our core wishes and desires, our hopes and dreams, our plans and expectations about our lives, and our future as well. The greatest threat we face comes from not embracing our own potential. Now is the time to hope, to fight, to dream, to unleash a new era of freedom, opportunity, and liberation.

Notes

Introduction

1. Kenny Torrella, "How will we feed Earth's rising population? Ask the Dutch," March 23, 2023, https://www.vox.com/.
2. Merrit Kennedy, "Tractor Trails of Protesting Dutch Farmers Snarl Traffic for Hundreds of Miles," NPR, October 1, 2019, https://www.npr.org/.
3. Reuters, "Dutch shelve billion-euro projects as EU nitrogen rules bite," September 13, 2019, https://www.reuters.com/.
4. Water Resources Mission Area, "Nutrients and Eutrophication," United States Geological Survey, March 3, 2019. https://www.usgs.gov/.
5. George Tanber, "Toxin leaves 500,000 in northwest Ohio without drinking water," Reuters, August 2, 2014. https://www.reuters.com/.
6. For a gripping account of the tragedy and its aftermath, see Scott Carney and Jason Miklian, *The Vortex: A True Story of History's Deadliest Storm, An Unspeakable War, and Liberation* (New York: Ecco Press, 2022).
7. Sue C. Grady, Joseph P. Messina, and Paul F. McCord, "Population Vulnerability and Disability in Kenya's Tsetse Fly Habitats," *PLOS Neglected Tropical Diseases* 5 (2011): e957.
8. Jeremy Youde, *Biopolitical Surveillance and Public Health in International Politics* (New York: Palgrave Macmillan, 2010), 1, 67.
9. Carrie Kahn, "Mexico City Keeps Sinking as Its Water Supply Wastes Away," NPR, September 14, 2018. https://www.npr.org/.
10. Hannah Beech, "How to Move a Sinking Capital City," *New York Times*, May 16, 2023.

11. Joseph Winters, "Rich countries are illegally exporting plastic trash to poor countries, data suggests," *Grist*, April 15, 2022, https://grist.org/.
12. Gilles Fauconnier and Mark Turner, *The Way We Think* (New York: Basic Books, 2002), 187.
13. Sebastian Wagner and Eduardo Zorita, "High-Resolution Climate Reconstruction of the Last 2,000 Years," in *Climate Changes in the Holocene*, ed. Eustathios Chiotis (Boca Raton, FL: CRC Press, 2019), 122.
14. See James Scott, *Against the Grain* (New Haven: Yale University Press, 2017).
15. Robert L. Lehrman, "Energy Is Not the Ability to Do Work." *The Physics Teacher* 11/1 (1973).
16. Larry Kirkpatrick and Gregory E. Francis, *Physics: A Conceptual Worldview* (Boston: Cengage Learning, 2009), 124
17. Atkins, *Four Laws That Drive the Universe*, 23.
18. William Thomson, "On a Universal Tendency in Nature to the Dissipation of Mechanical Energy," *Proceedings of the Royal Society of Edinburgh*, vol. 3 (Edinburgh: Neill and Company, 1857), 139–42.
19. Douglas C. Giancoli, *Physics for Scientists and Engineers* (London: Pearson Education, 2008), 545.
20. John M. Seddon and Julian D. Gale, *Thermodynamics and Statistical Mechanics* (London: Royal Society of Chemistry, 2001), 60–65.
21. Seddon and Gale, *Thermodynamics and Statistical Mechanics*, 65.
22. Atkins, *Four Laws That Drive the Universe*, 53.
23. For the famous reciprocal relations that describe heat flows, see Lars Onsager, "Reciprocal Relations in Irreversible Processes I," *Physical Review Journals* 37 (1931): 405–26. This work was the main reason why Onsager won the Nobel Prize in Chemistry. For a study of bosonic quantum gases in a one-dimensional trap, see Miguel Ángel García-March et al., "Non-equilibrium thermodynamics of harmonically trapped bosons," *New Journal of Physics* 18 (2016): 103035. For an exhaustive review of modern thermodynamics and an explanation of dissipative structures, which earned Ilya Prigogine his Nobel Prize, refer to Dilip Kondepudi and Ilya Prigogine, *Modern Thermodynamics: From Heat Engines to Dissipative Structures* (Hoboken, NJ: John Wiley & Sons, 2014), 421–41. In 2009, Alex Kleidon wrote an important theoretical study and review of the climate system using non-equilibrium thermodynamics. See Alex Kleidon, "Non-equilibrium thermodynamics and maximum entropy production in the Earth system," *Science of Nature* 96 (2009): 1–25.
24. A notable idea from the physicist Phil Attard looks at entropy as the number of particle configurations associated with a physical transition

in a given period of time. See Phil Attard, "The second entropy: A general theory for non-equilibrium thermodynamics and statistical mechanics," *Physical Chemistry* 105 (2009): 63–173. Perhaps the most technically rigorous model of entropy imagines it to be a collection of two functions that describe the changes happening among a restricted class of non-equilibrium systems. See Elliott H. Lieb and Jakob Yngvason, "The entropy concept for non-equilibrium states," *Proceedings of the Royal Society* 469 (2013): 1–15. The physicist Karo Michaelian provided an intuitive definition of entropy, viewing it as the rate at which physical systems explore available energy microstates. See K. Michaelian, "Thermodynamic dissipation theory for the origin of life," *Earth System Dynamics* (2011): 37–51.

25. Erwin Schrödinger, *What Is Life? The Physical Aspect of the Living Cell* (Ann Arbor: University of Michigan Press, 1945), 35–65.
26. Natalie Wolchover, "A New Physics Theory of Life," *Quanta Magazine*, January 22, 2014.
27. Carsten Hermann-Pillath, "Energy, growth, and evolution: Towards a naturalistic ontology of economics," *Ecological Economics* 119 (2015): 432–42.
28. Numerous studies from around the world have revealed a powerful relationship between energy use and economic growth. For a review of the statistical relationship between energy use and GDP growth worldwide, see Rögnvaldur Hannesson, "Energy and GDP growth," *International Journal of Energy Management*, vol. 3 (2009): 157–70. For a major study on the causality between energy and income in certain Asian countries, see John Asafu-Adjaye, "The relationship between energy consumption, energy prices, and economic growth: Time series evidence from Asian developing countries," *Energy Economics* 22 (2000): 615–25. For a general overview of how energy use has shaped human history, see Vaclav Smil, *Energy and Civilization* (Cambridge, MA: MIT Press, 2017).
29. Vaclav Smil, *Energy in Nature and Society: General Energetics of Complex Systems* (Cambridge, MA: MIT Press, 2008), 147–49.
30. Jerry H. Bentley, "Environmental Crises in World History," *Social and Behavioral Sciences* 77 (2013): 108–15.
31. Ibid., 113.
32. Will Steffen et al., "Global Change and the Earth System: A Planet Under Pressure," International Geosphere-Biosphere Programme (Stockholm: Royal Swedish Academy of Sciences, 2004), 2.
33. To pick just one of these, for an example of how global warming is increasing the odds of wildfires in the United States, see John T. Abatzoglou and A. Park Williams, "Impact of anthropogenic climate

change on wildfire across western US forests," *Earth, Atmospheric, and Planetary Sciences* 113 (2016): 11770–11775.
34. Karn Vohra et al., "Global mortality from outdoor fine particle pollution generated by fossil fuel combustion: Results from GEOS-Chem," *Environmental Research* 195 (2021): 110754.
35. An insightful explanation of these debates can be found in Carey King, *The Economic Superorganism* (New York: Springer International Publishing, 2020).

1. Growth and Scale in Economics

1. Senate Committee on Finance, *Revenue Proposals in the President's Fiscal Year 2006 Budget*, 2005, 42.
2. Ben Leubsdorf and Andrew Van Dam, "Steven Mnuchin Says U.S. Growth Can Be 3% to 4%. Here's Why That's Hard," *Wall Street Journal*, 2016.
3. Michael Mandelbaum, "A Bad Deal for America's Future," in *Project Syndicate*, 2011.
4. Organization for Economic Development and Education, *OECD Health Statistics 2022: Definitions, Sources, and Methods*, July 2022, https://www.oecd.org/.
5. Tim Callen, "Gross Domestic Product: An Economy's All," International Monetary Fund, 2012, https://www.imf.org/en/Home.
6. N. Gregory Mankiw, *Principles of Economics*, 9th ed. (Boston: Cengage, 2021), 470.
7. Annalisa Merelli, "Drugs, prostitution and smuggling—Italy's GDP is about to get much bigger," *Quartz*, May 26, 2014, https://qz.com/.
8. Katie Allen, "Accounting for drugs and prostitution to help push UK economy up by £65bn," *The Guardian*, June 10, 2014, https://www.theguardian.com/uk.
9. Reuters, "France keeps illegal drugs, prostitution out of economic output data," June 18, 2014, https://www.reuters.com/.
10. Matthias Schmelzer, *The Hegemony of Growth* (Cambridge: Cambridge University Press, 2016), 97.
11. OECD, "GDP per capita," in *National Accounts at a Glance* (Paris: OECD Publishing, 2013). The OECD states that "Gross Domestic Product (GDP) per capita is a core indicator of economic performance and commonly used as a broad measure of living standards or economic well-being, despite recognized shortcomings."
12. Mickey Francis, "U.S. energy intensity has dropped by half since 1983, varying greatly by state," U.S. Energy Information Administration, August 3, 2021, https://www.eia.gov/.
13. Christian Bogmans et al., "Energy, Efficiency Gains, and Economic

Development: When Will Global Energy Demand Saturate?," IMF Working Paper, 2020, https://www.imf.org/en/Home.
14. Our World in Data, "Carbon emission intensity of economies," 2018, https://ourworldindata.org/.
15. Elizabeth Gibney, "Largest Overhaul of Scientific Units Since 1875 Wins Approval," *Nature*, November 16, 2018, https://www.nature.com/.
16. National Geographic, "Meter Defined," *National Geographic*, October 19, 2023. https://www.nationalgeographic.org/society/.
17. For an excellent introduction to the aggregation problem, see Blair Fix, "The Aggregation Problem: Implications for Ecological and Biophysical Economics," *BioPhysical Economics and Resource Quality* 4 (2019). For a more technical treatment, see Jesus Felipe and Franklin Fisher, "Aggregation in Production Functions: What Applied Economists Should Know," *Metroeconomica* 54 (2003): 208–62.
18. Robert M. Solow, "A Contribution to the Theory of Economic Growth," *Quarterly Journal of Economics* 70 (1956): 65–94.
19. William Lazonick, "Profits Without Prosperity," *Harvard Business Review*, September 2014.
20. Doug Henwood, "The CHIPS Act Is a Massive Giveaway to Tech Companies," *Jacobin*, 2022, https://www.jacobin.com/.
21. Uri Friedman, "How Nigeria Became Africa's Largest Economy Overnight," *The Atlantic*, April 7, 2014, https://www.theatlantic.com/.
22. Blair Fix, "The aggregation problem: Implications for ecological and biophysical economics," *BioPhysical Economics and Resource Quality* 4 (2019).
23. See BEA News Release, "Gross Domestic Product, First Quarter 2024," Bureau of Economic Analysis, April 25, 2024, https://www.bea.gov/index.php/news/2024/gross-domestic-product-first-quarter-2024-advance-estimate.
24. Charles Steindel, "Chain-Weighting: The New Approach to Measuring GDP," *Current Issues in Economics and Finance* 1/9 (December 1995).
25. Bureau of Economic Analysis, *NIPA Handbook, Estimating Methods* (2022), chap. 4, 17–18.
26. For a concise and illuminating description of the method, see Michael Pettis, "Do We Understand the Math Behind the PPP Calculations?," Carnegie Endowment for International Peace, May 2, 2014, https://carnegieendowment.org/.
27. For the figures on steel, see World Steel Association, "Global crude steel output decreases by 0.9% in 2020," World Steel Association, January 25, 2021, https://worldsteel.org/. For the solar panel numbers, see Sayumi Take, "China's solar panel supply chain domination cause for worry:

IEA," *Nikkei*, July 7, 2022, https://asia.nikkei.com/. For the EV battery numbers, refer to Srinivas Mazumdaru, "EV Batteries: Can the West Catch Up with China?," *Deutsche Welle*, April 17, 2023, https://www.dw.com/.
28. An exajoule equals 10^{18} joules. See the *2022 Statistical Review of World Energy*, British Petroleum, 8.
29. Mary Hui, "China's life expectancy is now higher than that of the US," *Quartz*, September 1, 2022.
30. See Karl Marx, *Capital*, vol. 1 (London: Penguin, 1976).
31. See, for example, David Stern, "The role of energy in economic growth," *Ecological Economics Reviews* vol. 1219 1, February 2011, 26-51.
32. Carey King, *The Economic Superorganism* (New York: Springer International Publishing, 2020), 216–17.
33. Gregory Brunk, "Why Do Societies Collapse? A Theory Based on Self-Organized Criticality," *Journal of Theoretical Politics* 14 (2002): 195–230.
34. Ronald Bailey, *The End of Doom* (New York: St Martin's Press, 2015), 71.
35. Vaclav Smil, *Growth: From Microorganisms to Megacities* (Cambridge, MA: MIT Press, 2019), 509.
36. Ibid., 513.

2. The Neoclassical World

1. See Philip Mirowski, *More Heat Than Light* (Cambridge: Cambridge University Press, 1989).
2. Irving Fisher, *Mathematical Investigations in the Theory of Values and Prices*, from *Transactions of the Connecticut Academy*, vol. 9, July 1892, 85.
3. For an excellent discussion and refutation of Paul Samuelson's theory, refer to Stanley Wong, *Foundations of Paul Samuelson's Revealed Preference Theory* (Oxfordshire: Routledge & K. Paul, 1978).
4. For some examples, see Assar Lindbeck and Jörgen W. Weibull, "Altruism and Time Consistency: The Economics of Fait Accompli," *Journal of Political Economy* 96 (1988): 1165–82. Also see Klaus Jaffe, "An economic analysis of altruism: Who benefits from altruistic acts?," *Journal of Artificial Societies and Social Simulation* 5 (2022). For criticism of these kinds of approaches, see Kristen Renwick Monroe, "A Fat Lady in a Corset: Altruism and Social Theory," *American Journal of Political Science* 38 (1994): 861–93. Another notable critique is from Herbert A. Simon, "Altruism and Economics," *American Economic Review* 83/2 (1993): 155–61.
5. See Richard Thaler, *Misbehaving: The Making of Behavioral Economics*

NOTES TO PAGES 76 – 84   359

(New York: W. W. Norton, 2016). Also refer to Daniel Kahneman, *Thinking, Fast and Slow* (New York: Farrar, Straus and Giroux, 2011).

6. The economist Joan Robinson was one of the most brilliant critics of neoclassical economics. For her classic critique of utility and neoclassical theory more broadly, see Joan Robinson, *Economic Philosophy* (Middlesex, UK: Penguin, 1964).

7. The economist Larry Summers was perhaps the worst offender. For an example of his usual pablum, see Alvin Powell, "Summers says pandemic only partly to blame for record inflation," *Harvard Gazette*, February 4, 2022, https://news.harvard.edu/gazette/.

8. Contrary to the assertions of Larry Summers, for example, much of the post-pandemic inflationary wave was driven by dominant corporations using a moment of historical crisis to jack up prices. See Tom Perkins, "Top US corporations raising prices on Americans even as profits surge," *The Guardian*, April 27, 2022, https://www.theguardian.com/us. For an excellent academic approach on the subject, see Isabella Weber Evan Wasner, "Sellers' inflation, profits, and conflict: Why can large firms hike prices in an emergency?" *Review of Keynesian Economics* 11 (2023): 183–213. For broader discussions about the relationship between prices and sociopolitical dynamics, see Jonathan Nitzan, "Inflation as restructuring: A theoretical and empirical account of the U.S. experience," PhD dissertation, McGill University, 1992. Also see Jonathan Nitzan and Shimshon Bichler, *Capital as Power* (New York: Routledge, 2009).

9. John Emshwiller and Neil Behrmann, "Restored Luster: How De Beers Revived World Diamond Cartel After Zaire's Pullout," *Wall Street Journal*, July 7, 1983.

10. For example, see K. Rajagopalachar, *Business Economics* (New Delhi: Atlantic Publishers, 1993), 105–6.

11. See Susan Feigenbaum, *Principles of Macroeconomics: The Way We Live* (New York: Macmillan, 2011), 519, G-1 Glossary.

12. See M. Kasi Reddy and S. Saraswathi, *Managerial Economics and Financial Accounting*, 2007, 47

13. For an excellent description of these results, see S. Abu Turab Rizvi, "The Sonnenschein-Mantel-Debreu Results After Thirty Years," *History of Political Economy* 38 (2006): 228–45.

14. In addition to the review paper from Rizvi cited above, another great review paper on the subject is Frank Ackerman, "Still Dead After All These Years: Interpreting the Failure of General Equilibrium Theory," *Journal of Economic Methodology* (2002): 119–39. For the individual economists mentioned in the paragraph, see Alan Kirman, "The intrinsic limits of modern economic theory: the emperor has no

clothes," *Economic Journal* 395 (1989): 126–39. Next, see Donald Saari, "Mathematical complexity of simple economics," *Notices of the AMS* 42 (1995). Finally, see Donald Brown and Chris Shannon, "Uniqueness, stability, and comparative statics in rationalizable Walrasian markets," *Econometrica* (2000): 1529–39.

15. For the quote itself, refer to the highly misguided article by Samuel Brittan, "Thatcher was right—there is no society," *Financial Times*, April 18, 2013.
16. Mariana Mazzucato, *The Value of Everything* (New York: Hachette, 2018).
17. John Bates Clark, *The Distribution of Wealth: A Theory of Wages, Interest and Profits*, Preface, Econlib Books, https://www.econlib.org/.
18. Blair Fix, "No, Productivity Does Not Explain Income," *Capital as Power,* January 17, 2020, https://capitalaspower.com/.
19. For an excellent summary, see Blair Fix, "The trouble with human capital theory," Working Paper, https://capitalaspower.com, 2018. Also see John Hunter and Frank Schmidt, "Individual differences in output variability as a function of job complexity," *Journal of Applied Psychology* 75 (1990): 28–42.
20. For just one example, see Prasert Kanawattanachai and Youngjin Yoo, "The Impact of Knowledge Coordination on Virtual Team Performance over Time," *MIS Quarterly* 31 (2007): 783–08.
21. For an excellent explanation of the theory, see Robert Murphy, "Böhm-Bawerk's Critique of the Exploitation Theory of Interest," Mises Institute, November 26, 2004, https://mises.org/.
22. See Rena S. Miller and Gary Shorter, "High-Frequency Trading: Overview of Recent Developments," *Congressional Research Service*, April 4, 2016, https://www.everycrsreport.com/. Also see Matteo Aquilina, Eric Budish, Peter O'Neill, "Quantifying the High-Frequency Trading Arms Race," *Quarterly Journal of Economics* 137 (2022): 493–564.
23. For example, see Igor Makarov and Antoinette Schoar, "Trading and arbitrage in cryptocurrency markets," *Journal of Financial Economics* 135 (2020): 293–319. See also Craig Burnside, Martin Eichenbaum, and Sergio Rebelo, "Carry Trade and Momentum in Currency Markets," *Annual Review of Financial Economics* 3 (2011), 511–35.
24. Federal Reserve, press release, "Federal Reserve Board issues final rule regarding dividend payments on Reserve Bank capital stock," November 23, 2016, https://www.federalreserve.gov/.

3. Theories of Economic Growth and Development

1. See Robert M. Solow, "A Contribution to the Theory of Economic Growth," *Quarterly Journal of Economics* 70 (1956): 65–94.

2. Solow's conclusions relied on a function known as the Cobb-Douglas production function, which was widely used in economics back then and remains popular to this very day. However, Anwar Shaikh brilliantly tore apart Solow's results in 1974, when he showed that the Cobb-Douglas production function could accurately model data sets for which a production function should not exist. See Anwar Shaikh, "Laws of Production and Laws of Algebra: The Humbug Production Function," *Review of Economics and Statistics* 56 (1974): 115–20. It turns out that there is a kind of trivial and profound reason for this behavior. In 2005, Jesus Felipe and J. S. L. McCombie convincingly proved what had been widely understood by Shaikh and others: the Cobb-Douglas production function is nothing more than an identity equation, a different way of writing the additive equation *capital plus labor*, and it reveals absolutely nothing about the neoclassical theory of distribution. See Jesus Felipe and J. S. L. McCombie, "How Sound Are the Foundations of the Aggregate Production Function?," *Eastern Economic Journal* 31 (2005): 467–88. The Cobb-Douglas function was picking up the factor shares inherent in the empirical data sets, most of which had roughly constant factor shares, because its exponents are exactly the same as the factor shares from the identity equation. In sum, the Cobb-Douglas production function is an elaborate way of saying that one equals one.
3. See David Stern, "Economic Growth and Energy," in *Encyclopedia of Energy*, ed. C. J. Cleveland (San Diego: Academic Press, 2004), 40.
4. For an excellent description and critique of DICE, see Steve Keen, "The appallingly bad neoclassical economics of climate change," *Economics and Climate Emergency* 18 (2021): 1149–77.
5. William Nordhaus, "Revisiting the social cost of carbon," *Earth, Atmospheric, and Planetary Sciences* 114 (2017): 1518–23.
6. For a good explanation of what Nordhaus did, refer again to Steve Keen, "The appallingly bad neoclassical economics of climate change," *Economics and Climate Emergency* 18 (2021): 1149–77.
7. The paper in question is Timothy Lenton et al., "Tipping elements in the Earth's climate system," *Proceedings of the National Academy of Sciences* 105 (2008): 1786–93.
8. Ibid.
9. Steve Keen, "The appallingly bad neoclassical economics of climate change," *Economics and Climate Emergency* 18 (2021): 1149–77.
10. See Robert Pindyck, "Climate change policy: What do the models tell us?," *Journal of Economic Literature* 51 (2013): 860-872.
11. The Carnot cycle is the reversible thermodynamic cycle of the most efficient classical thermodynamic engine possible. See John W. Jewett

and Raymond A. Serway, *Physics for Scientists and Engineers* (Boston: Cengage Learning, 2008), 618.

12. Stanley Reed, "Germany's Shift to Green Power Stalls, Despite Huge Investments," *New York Times*, October 7, 2017.
13. Tobias Buck, "Energy Shift Fails to Cut German Carbon," *Financial Times*, October 8, 2018.
14. For a comprehensive guide to some recent research on hurricanes and climate change, see Jennifer M. Collins and Kevin Walsh, eds., *Hurricanes and Climate Change*, vol. 3 (New York: Springer, 2017). For a review of the role climate change plays in the spread of infectious diseases, see Xiaoxu Wu et al., "Impact of Climate Change on Human Infectious Diseases: Empirical Evidence and Human Adaption," *Environment International* 86 (2016): 14–23.
15. See, for example, Jerry H. Bentley, "Environmental Crises in World History," *Procedia—Social and Behavioral Sciences* 77 (2013): 108–15.
16. For more on Marx and his theory of the metabolic rift, see John Bellamy Foster, *Marx's Ecology* (New York: Monthly Review Press, 2000).
17. See Johan Rockström et al., "A Safe Operating Space for Humanity," *Nature* 461 (2009): 472–75.
18. Stephen E. Kesler, Adam C. Simon, and Adam F. Simon, *Mineral Resources, Economics and the Environment* (Cambridge: Cambridge University Press, 2015), 302.
19. Anne Davies and Ben Doherty, "Corruption, Incompetence and a Musical: Nauru's Cursed History," *The Guardian*, September 3, 2018.
20. For an excellent introduction to how photovoltaics work, see "Solar Cell Efficiency," *Energy Education*, June 25, 2018.
21. Alexis De Vos, "Detailed Balance Limit of the Efficiency of Tandem Solar Cells," *Journal of Physics D: Applied Physics* 13 (1980): 839–46.
22. See, for example, Tim Worstall, "When Physicists Do Economics We Seem Not to Get Economics as the Result," *Forbes*, October 6, 2014.
23. David Stern, "Economic Growth and Energy," in *Encyclopedia of Energy*, ed. C. J. Cleveland (San Diego: Academic Press, 2004), 42.
24. See Robert M. Solow, "The Economics of Resources or the Resources of Economics," *American Economic Review* 64 (1974): 1–14.
25. See Annalies Winny, "Life Expectancy Is Declining in the U.S.—It Doesn't Have to Be," *Johnc Hopkins School of Public Health*, December 6, 2022, https://publichealth.jhu.edu/.
26. See Hannah Ritchie, "Many countries have decoupled economic growth from CO_2 emissions, even if we take offshored production into account," *Our World in Data*, December 1, 2021, https://ourworldindata.org/.
27. I confirmed this in an email exchange with the Global Carbon Project.

NOTES TO PAGES 118 – 123 363

Their reply to my question about whether they measured fluorinated gases was "We don't do synthetic gases unfortunately."
28. Russell Hotten, "Volkswagen: The Scandal Explained," BBC, 2015, https://www.bbc.com/.
29. See Mark Brownstein, Steven Hamburg, and Ramon Alvarez, "Major studies reveal 60% more methane emissions," from the Environmental Defense Fund, 2018.
30. See "Methane emissions from the energy sector are 70% higher than official figures" from the International Energy Agency, 2022. The IEA publicizes global methane emissions data through its Global Methane Tracker report.
31. NASA Staff, "Methane," NASA, 2022, https://climate.nasa.gov/.
32. Chris Mooney et al., "Countries' climate pledged built on flawed data," *Washington Post*, November 7, 2021.
33. Don Carrington, "Analysts Call Apple Renewable Energy Claims 'Lies,'" *Carolina Journal*, December 2, 2015.
34. Ibid.
35. Ella Fanger, "The Dirty Energy Fueling Amazon's Data Gold Rush," *The Nation*, February 22, 2024.
36. For example, see the Net-Zero Banking Alliance working with the United Nations. The Alliance has over a hundred members, including some major American powerhouses like JPMorgan and Citi.
37. See the 2022 report "Banking on Climate Chaos" from the Sierra Club, Rainforest Action Network, Indigenous Environmental Network, and other organizations.
38. Olivia Rosane, "Carbon Offsets 101: Why We Can't Offset Our Way Out of the Climate Crisis," *EcoWatch*, October 20, 2022, https://www.ecowatch.com/.
39. Lisa Song and Paula Moura, "An Even More Inconvenient Truth," *ProPublica*, May 22, 2019, https://www.propublica.org/
40. For an excellent description of the issues at play with the extraction of palm oil in Indonesia, see Abrahm Lustgarten, "Palm Oil Was Supposed to Help Save the Planet. Instead, It Unleashed Catastrophe," *New York Times*, November 20, 2018.
41. *U.S. Energy Information Administration*, see Table 1.7 in "Primary Energy Consumption, Energy Expenditures, and Carbon Dioxide Emissions Indicators," March 26, 2019.
42. For some relatively recent flavors of elite opinion on this topic, see John L. Seitz and Kristen A. Hite, *Global Issues: An Introduction* (Hoboken, NJ: John Wiley & Sons, 2012), 126. Also look at Devashree Saha and Mark Muro, "Growth, Carbon, and Trump: States Are 'Decoupling' Economic Growth from Emissions Growth," *Brookings Institution*, December 8, 2016.

43. See David Stern, "Economic Growth and Energy," in *Encyclopedia of Energy*, ed. C. J. Cleveland (San Diego: Academic Press, 2004), 35–51. Kaufmann also documented the structural shifts that occurred within the U.S. energy sector and analyzed their impact on economic growth. See Robert Kaufmann, "The Mechanisms for Autonomous Energy Efficiency Increases: A Cointegration Analysis of the US Energy/GDP Ratio," *Energy Journal* 25 (2004): 63–86.
44. Zeke Hausfather, "Analysis: Why US Carbon Emissions Have Fallen 14% Since 2005," *Carbon Brief*, August 15, 2017.
45. Brad Plumer, "U.S. Carbon Emissions Surged in 2018 Even as Coal Plants Closed," *New York Times*, January 8, 2019.
46. See, for example, the report "CO2 Emissions in 2022" from the International Energy Agency, March 2023.
47. For one of the most influential studies in this field, see David I. Stern, "The Role of Energy in Economic Growth," *Crawford School Centre for Climate Economics & Policy, Paper No. 3.10* (2011). For a review of the statistical relationship between energy use and GDP growth worldwide, see Rögnvaldur Hannesson, "Energy and GDP Growth," *International Journal of Energy Management* 3 (2009): 157–70. For a major study on the link between energy and income in certain Asian countries, see John Asafu-Adjaye, "The Relationship between Energy Consumption, Energy Prices, and Economic Growth: Time Series Evidence from Asian Developing Countries," *Energy Economics* 22 (2000): 615–25. For a general overview of how energy use has shaped human history, see Vaclav Smil, *Energy and Civilization* (Cambridge, MA: MIT Press, 2017).
48. Jessica Dickler, "Consumer Debt Hits $4 Trillion," *CNBC*, February 21, 2019.
49. Drew Desilver, "For Most U.S. Workers, Real Wages Have Barely Budged in Decades," *Pew Research Center*, August 7, 2018.
50. See the report "Decoupling of Global Emissions and Economic Growth Confirmed," *International Energy Agency*, March 16, 2016.
51. Zeke Hausfather, "Analysis: Global CO2 Emissions Set to Rise 2% in 2017 After Three-Year Plateau," *Carbon Brief*, November 13, 2017.
52. Damian Carrington, "Brutal News: Global Carbon Emissions Jump to All-Time High in 2018," *The Guardian*, December 5, 2018.
53. James Ward et al., "The Decoupling Delusion: Rethinking Growth and Sustainability," *The Conversation*, March 12, 2017. https://theconversation.com/us.
54. See Goher Ur Rehman Mir and Servaas Storm, "Carbon Emissions and Economic Growth: Production-based versus Consumption-based Evidence on Decoupling," Institute for New Economic Thinking, April

NOTES TO PAGES 127 – 144

2016. Also see Dagmawe Tenaw and Alemu L. Hawitibo, "Carbon decoupling and economic growth in Africa: Evidence from production and consumption-based carbon emissions," *Resources, Environment and Sustainability* 6 (2021): 100040.
55. See Joel Mokyr, *The Culture of Growth: The Origins of the Modern Economy* (Princeton: Princeton University Press, 2016).
56. Ibid., 8.
57. Ibid.
58. Daron Acemoglu and James Robinson, *Why Nations Fail* (New York: Crown Business, 2012), 73.
59. Ibid., 76–78.
60. Karl Marx and Friedrich Engels, *The Communist Manifesto*, ed. Martin Malia (New York: Penguin, 1998), 57.
61. Acemoglu and Robinson, *Why Nations Fail*, 79.
62. Ibid., 208.
63. For Acemoglu and Robinson's overview of Venetian economic and political history in the Middle Ages, see pp. 152–58 of *Why Nations Fail*.
64. See Jason Roche, "Consequences," in *The Crusades to the Holy Land*, ed. Alan Murray (Santa Barbara: ABC-CLIO, 2015).
65. See "Population of the Venetian Republic in 1557, by region" from the Statista Research Department, https://www.statista.com/. According to Statista researchers: "By the mid-16th century, the population of the Venetian Republic was roughly 2.3 million people, at a time when Europe's population was around 70 million. 1.7 million of this population was concentrated in northeast Italy, while the islands of Crete and Cyprus were the most populous overseas territories." For England's population figures, see the chart "The Population of England" from the public data repository Our World in Data, https://ourworldindata.org/.
66. Robert Allen, *The British Industrial Revolution in Global Perspective* (Cambridge: Cambridge University Press, 2009), 129–30.
67. Acemoglu and Robinson, *Why Nations Fail*, 185.
68. Allen, *The British Industrial Revolution*, 129–30.
69. Ibid..
70. Acemoglu and Robinson, *Why Nations Fail*, 54–55.
71. See Jared Diamond, *Guns, Germs, and Steel* (New York: W. W. Norton, 1997), 16.
72. For some prominent papers on the subject, see Harvey Weiss and Raymond Bradley, "What drives societal collapse?," *Science* 291/5504 (2001), as well as Peter Douglas et al., "Impacts of Climate Change on the Collapse of Lowland Maya Civilization," *Annual Review of Earth*

and Planetary Sciences 44 (2016: 613–45). For some of the more notable English-language books on the subject, see Brian Fagan, *The Great Warming: Climate change and the rise and fall of civilizations* (London: Bloomsbury Publishing, 2008); and Jared Diamond, *Collapse* (New York: Penguin, 2005).

4. Bionomic Disruption and the Future of Humanity

1. For a short description of the process, see Kartik Aiyer, "The Great Oxidation Event: How Cyanobacteria Changed Life," *American Society for Microbiology*, February 18, 2022.
2. See Jason W. Moore, *Anthropocene or Capitalocene? Nature, History, and the Crisis of Capitalism* (Oakland, CA: PM Press, 2016). Also see John Bellamy Foster and Brett Clark, "The Capitalinian: The First Geological Age of the Anthropocene," *Monthly Review* 73/ 4 (September 1, 2021).
3. Kyle Paoletta, "The Incredible Disappearing Doomsday," *Harper's Magazine*, April 2023.
4. See Paul Ehrlich, *The Population Bomb* (Oakland, CA: Sierra Club, 1968).
5. Immanuel Kant, *To Perpetual Peace: A Philosophical Sketch*, trans. Ted Humphrey (Indianapolis: Hackett Publishing Company, 2003).
6. Joseph Kane and Kirsten Lydic, "Basic Income Does Not Threaten Labor Markets," in *Political Activism and Basic Income Guarantee*, ed. Richard Kaputo and Larry Liu (London: Palgrave Macmillan, 2020), 57.
7. Waldemar Kaempffert, "Miracles You'll See in the Next 50 Years," *Popular Mechanics*, February 1950.
8. Philip Max Hartmann, *Profiting from Artificial Intelligence* (Paris: Books on Demand, 2020), 12.
9. John Hinshaw and Peter Stearns, *Industrialization in the Modern World* (Santa Barbara, CA: ABC-CLIO, 2013), 338.
10. Richard Brooks, *Atlas of World Military History* (London: HarperCollins, 2000), 142.
11. Bjørn Lomborg, *The Skeptical Environmentalist* (Cambridge: Cambridge University Press, 1998), 241.
12. For a prominent paper on the subject see Gerardo Ceballos et al., "Accelerated modern human-induced species losses: Entering the sixth mass extinction," *Science Advances* 1 (2015).
13. Robert Cowie, Philippe Bouchet, and Benoît Fontaine, "The Sixth Mass Extinction: Fact, fiction, or speculation?" *Biological Reviews* 97 (2022): 640–63.
14. Dave Goulson, "The insect apocalypse, and why it matters," *Current Biology* 29 (2019): 967–71.
15. Gregor Claus et al., "Challenges in Cocoa Pollination: The Case of

NOTES TO PAGES 154 – 158 367

Côte d'Ivoire," in *Pollination in Plants*, ed. Phatlane William Mokwala, (London: Intechopen, 2018).
16. Yinon Bar-On, Rob Phillips, and Ron Milo, "The biomass distribution on Earth," *Biological Sciences* 115 (2018): 6506–11.
17. Bjørn Lomborg, *The Skeptical Environmentalist* (Cambridge: Cambridge University Press, 2001), 535.
18. Ibid., 523.
19. See Patrick Kinney et al., "Winter season mortality: Will climate warming bring benefits?," *Environmental Research Letters* 10 (2015): 064016.
20. See Qi Zhao et al., "Global, regional, and national burden of mortality associated with non-optimal ambient temperatures from 2000 to 2019," *The Lancet Planetary Health* 5 (2021): 415–25.
21. Bjørn Lomborg, "Don't buy the latest climate change alarmism," *New York Post*, August 9, 2021.
22. Aaron Bernstein et al., "Global warming contributes to increased heat-related mortality," *Climate Feedback*, September 22, 2021.
23. Ibid.
24. See Bjørn Lomborg, "The heresy of heat and cold deaths," https://www.lomborg.com/.
25. See Bjørn Lomborg, "Welfare in the 21st century: Increasing development, reducing inequality, the impact of climate change, and the cost of climate policies," *Technological Forecasting and Social Change* 156 (2020). In this paper, Lomborg cites results from an earlier 2018 paper showing that "areas in severe drought" have marginally decreased over the past century. The categorization is bizarre as areas in "severe" drought don't necessarily imply anything about areas experiencing some kind of drought in general. What's worse, the chart shown by Lomborg is based on the Standard Precipitation Index (SPI), a flawed metric for measuring actual drought conditions. In reality, the Palmer Drought Severity Index (PDSI) is a superior metric because it considers precipitation, temperature, and soil moisture, in contrast to the SPI, which only uses precipitation as an input. The SPI therefore has the problem that it doesn't account for surface warming, so it assumes that an area with normal rainfall is not experiencing any drought even though extreme heat might be evaporating much of the water.
26. Aiguo Dai, "Increasing drought under global warming in observations and models," *Nature Climate Change* 3 (2013): 52–58.
27. Kevin Trenberth et al., "Global warming and changes in drought," *Nature Climate Change* 4 (2013): 17–22.
28. See Aiguo Dai, Kevin Trenberth, and Taotao Qian, "A global dataset of Palmer Drought Severity Index for 1870-2002: Relationship with soil

moisture and effects of surface warming," *Journal of Hydrometeorology* 5 (2004): 1117–30.
29. Karn Vohra et al., "Global mortality from outdoor fine particle pollution generated by fossil fuel combustion: Results from GEOS-Chem," *Environmental Research* 195 (2021).
30. Jos Lelieveld et al., "Loss of life expectancy from air pollution compared to other risk factors: a worldwide perspective," *Cardiovascular Research* 116 (2020): 1910–17.
31. Jos Lelieveld et al., "Effects of fossil fuel and total anthropogenic emission removal on public health and climate," *Earth, Atmospheric, and Planetary Sciences* 116 (2019): 7192–97.
32. International Energy Agency, "Energy and Air Pollution" (Paris: International Energy Agency, 2016), 13.
33. Ibid.
34. Ibid.
35. Ronald Bailey, *The End of Doom: Environmental Renewal in the Twenty-first Century* (New York: Thomas Dunne Books, 2015), 71–73.
36. See Catherine Taylor et al., "Britain used to rule a quarter of the world. What happened?" Australian Broadcasting Corporation, September 17, 2022.
37. Bailey, *The End of Doom*, 68.
38. For example, see Timothy Kohler et al., "Greater post-Neolithic wealth disparities in Eurasia than in North America and Mesoamerica," *Nature* 551 (2017): 619–22.
39. See Joseph Inikori, *Africans and the Industrial Revolution in England: A Study in International Trade and Economic Development* (Cambridge: Cambridge University Press, 2002).
40. See, for example, Ronald Formisano, *Plutocracy in America* (Baltimore: Johns Hopkins University Press, 2015); and Thomas Ferguson, *Golden Rule: The Investment Theory of Party Competition and the Logic of Money-Driven Political Systems* (Chicago: Chicago University Press, 1995).
41. Jessica Whyte, *The Morals of the Market* (New York: Verso Books, 2019), 54.
42. See Delegate Alert, "Basel Plastic Waste Trade Violations Rampant One Year After Amendments Entry into Force," Basel Action Network, February 25, 2022. Also see Joseph Winters, "Rich countries export twice as much plastic waste to the developing world as previously thought," *Grist*, March 13, 2023. https://grist.org/.
43. See Madeleine Zelin, "The Grandeur of the Qing Economy," Asian Topics in World History, Columbia University, 2005. https://projects.mcah.columbia.edu/nanxuntu/html/economy/.

44. Amitav Ghosh, *The Nutmeg's Curse* (Chicago: University of Chicago Press, 2022), 42.
45. For a thorough critique of the school, refer to Sebastian Rosato, "The Flawed Logic of Democratic Peace Theory," *American Political Science Review* 97 (2003): 585–602.
46. Steven Pinker, *The Better Angels of Our Nature* (New York: Viking Books, 2011), 193.
47. Ibid., 195.
48. Valerie Hansen, *The Open Empire* (New York: W. W. Norton, 2000), 228.
49. Richard Guisso, *Cambridge History of China* (Cambridge: Cambridge University Press, 1979), 297.
50. Michael White, *The Great Big Book of Horrible Things* (New York: W. W. Norton, 2011), 38–40.
51. Richard Holmes, *World War II: The Definitive Visual History* (New York: DK Publishing, 2009), 334.
52. The Conflict Catalog also claims that there were only two sides, the French and the English, during this period. The critical role played by the Scottish and the Burgundians in this period is completely ignored. And to clarify further: the "Hundred Years' War" actually lasted more than 100 years and overall serves as just a lazy label for numerous major wars and conflicts that unfolded between various European powers, centered around the French and the English, from 1337 to 1453. One may also think that per capita death tolls from war can be simply calculated for every single distinct year of recorded human history, regardless of what wars were happening in that year. It's true that this method would remove the aggregation problem identified here, but in practice it's almost impossible to pull off in any reliable way. That's because you'd have to go around the world in every year, find all the conflicts and battles that took place in that year, estimate the deaths from all those conflicts and battles, then aggregate everything together. Given the paucity of source data and the practical futility of this exercise, most people just aggregate by grouping deaths together through distinct wars, thus leaving open the problem of what counts as a war and what doesn't.
53. Yinon Bar-On, Rob Phillips, and Ron Milo, "The biomass distribution on Earth," *Biological Sciences* 115 (2018): 6506–11.
54. Vaclav Smil, "Harvesting the Biosphere: The Human Impact," *Population and Development Review* 37 (2011): 613–36.
55. See Yude Pan et al., "The Structure, Distribution and Biomass of the World's Forests," *Annual Review of Ecology, Evolution, and Systematics* 44 (2013): 593–622. A hectare is 10,000 square meters, a square with 100 meters on each side, or roughly 107,000 square feet.

56. Xiao-Peng Song et al., "Global land change from 1982 to 2016," *Nature* 560 (2018): 639–43.
57. See David Gibbs, "By the Numbers: The Value of Tropical Forests in the Climate Change Equation," Global Forest Watch, October 5, 2019.
58. See Johan Rockström et al., "A Safe Operating Space for Humanity," *Nature* 461 (2009): 472–75.
59. Nicolás Misculin and Gabriel Burin, "How a Year of 'Endless Storms' Battered Argentina's Economy," Reuters, December 20, 2018.
60. Adam Wernick, "Climate Change Is the Overlooked Driver of Central American Migration," PRI, February 6, 2019.
61. Rupam Jain, "In Parched Afghanistan, Drought Sharpens Water Dispute with Iran," Reuters, July 16, 2018.
62. Keith Bradsher, "Drought Hurts China's Economy as Central Bank Cuts Rates," *New York Times*, August 22, 2022. Also see Keith Bradsher and Joy Dong, "China's Record Drought Is Drying Its Rivers and Feeding Its Coal Habit," *New York Times*, August 26, 2022.
63. Liz Alderman, "French Nuclear Power Crisis Frustrates Europe's Push to Quit Russian Energy," *New York Times*, June 19, 2022. Also see Kim Willsher, "Macron warns of 'end of abundance' as France faces difficult winter," *The Guardian*, August 24, 2022.
64. See Max Bearak, Raymond Zhong, and Ihsanullah Tipu Mehsud, "Deadly Floods Devastate an Already Fragile Pakistan," *New York Times*, August 29, 2022. For attribution to global warming, see Raymond Zhong, "In a First Study of Pakistan's Floods, Scientists See Climate Change at Work," *New York Times*, September 15, 2022.
65. Heinz Schandl et al., *Global Material Flows and Resource Productivity* (Paris: International Resource Panel, 2016), 40.
66. Frédéric Simon, "Decoupling Energy from GDP Growth 'Might Be Impossible,' Statoil Says," *Euractiv*, June 15, 2017. https://www.euractiv.com/.
67. Intergovernmental Panel on Climate Change, "Summary for Policymakers of IPCC Special Report on Global Warming," October 8, 2018. https://www.ipcc.ch/.
68. Brady Dennis and Chris Mooney, "We Are in Trouble: Global Carbon Emissions Reached a Record High in 2018," *Washington Post*, December 5, 2018.

5. Energetic Conversions in the Economic Process

1. See David Halliday and Robert Resnick, *Fundamentals of Physics* (Hoboken, NJ: John Wiley & Sons, 1988), 516.
2. Ibrahim Dincer and Marc A. Rosen, *Exergy: Energy, Environment and Sustainable Development* (Amsterdam: Elsevier Science, 2007), 73.

3. Murray Patterson, "What Is Energy Efficiency?," *Energy Policy* 24 (1996): 377–90.
4. For Patterson's proposed solution to the quality problem, see Murray Patterson, "Commensuration and theories of value in ecological economics," *Ecological Economics* 25 (1998): 105–25.
5. Matthew Daly, "Feds order review of power-grid security after attacks," Associated Press, December 15, 2022, https://apnews.com/.
6. See "Primary Energy Consumption," *Organization for Economic Cooperation and Development*, November 20, 2001. For a list of things included under primary consumption in the United States, see the entry "Primary Energy Consumption" in the Glossary of the U.S. Energy Information Administration, https://www.eia.gov/tools/glossary/.
7. For an explanation of these methods, see "Statistics questionnaires" by the International Energy Agency. https://www.iea.org/.
8. BP assumes an efficiency rating of 38 percent for the hypothetical power plant. See *BP Statistical Review of World Energy* (London: British Petroleum, 2018), 52.
9. See Charles A. S. Hall, *Energy Return on Investment* (Berlin: Springer Nature, 2017).
10. A BTU is equal to 1,055 joules, the standard unit of energy. A joule is roughly the amount of energy it takes to lift an apple up to your mouth. See "Average Operating Heat Rate for Selected Energy Sources," Table 8.1, *U.S. Energy Information Administration*, October 22, 2018.
11. For his primary work on the subject, see Vaclav Smil, *Power Density: A Key to Understanding Energy Sources and Uses* (Cambridge, MA: MIT Press, 2015).
12. Ibid., ix.
13. Arnulf Grubler, "Energy: Profiles of Power," *Nature* 523 (2015): 32–33.

6. The Ecodynamic Synthesis

1. Karl Marx, *Capital*, vol. 1 (London: Penguin, 1976), 283.
2. John Bellamy Foster and Brett Clark, *The Robbery of Nature* (New York: Monthly Review Press, 2020), 32–33.
3. Ibid., 36.
4. Jared Diamond, *Collapse* (New York: Viking Penguin, 2005), 286–93.
5. Ibid., 79–119.
6. For a recent example, see Valentí Rull, "The Deforestation of Easter Island," *Biological Reviews* 95 (2019): 124–41.
7. Lawrence Ekeh, *Industrialization and National Prosperity* (London: LUZEK Publishers, 2009), 8–9.
8. Kent Deng, *Mapping China's Growth and Development in the Long Run* (Singapore: World Scientific Publishing, 2015), 105–6.

9. Ibid.
10. Arnold Pacey, *Technology in World Civilization: A Thousand-Year History* (Cambridge. MA: MIT Press, 1991), 7.
11. Tim Wright, *Coal Mining in China's Economy and Society* (Cambridge: Cambridge University Press, 1984), 7–9.
12. Ibid.
13. Jared Diamond, *Collapse* (New York: Viking Penguin, 2005), 294–304.
14. Ibid., 302.
15. Ibid., 300.

7. Technological Dynamics of Growth and Stability

1. Shawn Miller, "Where's the Innovation? An Analysis of Quantities and Qualities of Anticipated Obvious Patents," *Virginia Journal of Law and Technology* 18 (2013).
2. See Donald Light and Rebecca Warburton, "Demythologizing the high costs of pharmaceutical research," *BioSocieties* 6 (2011): 34–50.
3. See Jesus Felipe and F. Gerard Adams, "The Estimation of the Cobb-Douglas Function: A Retrospective View," *Eastern Economic Journal* 31 (2005): 427–45.
4. Steven Pomeroy, *An Untaken Road* (Annapolis, MD: Naval Institute Press, 2016), 19.
5. Lynn White Jr., *Medieval Technology and Social Change* (London: Oxford University Press, 1962). See also Thomas Barnebeck Andersen, Peter Sandholt Jensen, Christian Volmar Skovsgaard, "The heavy plow and the agricultural revolution in Medieval Europe," *Journal of Development Economics* 118 (2016): 133–49.
6. Brian C. Black, *To Have and Have Not: Energy in World History* (Lanham, MD: Rowman & Littlefield, 2022), 110.
7. See Langdon Winner, "Artifice and Order," in *Technology and Values*, ed. Craig Hanks (Oxford: Wiley-Blackwell, 2010), 82.
8. See Humberto Maturana and Francisco Valero, *Autopoiesis and Cognition* (Dordrecht: D. Reidel Publishing Company, 1980).
9. Stuart Kauffman, "Autocatalytic sets of proteins," *Journal of Theoretical Biology* 119 (1986): 1–24.
10. IEA, *World Energy Outlook 2022* (Paris: IEA, 2022), 279, https://www.iea.org/.
11. The term "rare earth" element is misleading because many of these elements are not scarce at all. Some of them are far more abundant than things like copper and gold. They're called "rare earth" because they're not heavily concentrated in mineral ore deposits, unlike other metals. Instead, they're scattered around the Earth in smaller concentrations. The lanthanides are 15 metallic elements from

lanthanum to lutetium in the Periodic Table, with atomic numbers from 57 to 71.
12. IEA, *The Role of Critical Minerals in Clean Energy Transitions* (Paris: IEA, 2021), 5, https://www.iea.org/.
13. Ibid.
14. Ibid.
15. Brad Plumer, "The U.S. Has Billions for Wind and Solar Projects. Good Luck Plugging Them In," *New York Times*, February 23, 2023.
16. Ibid.
17. Tom Simonite, "Moore's Law Is Dead. Now What?" *MIT Technology Review*, May 13, 2016, https://www.technologyreview.com/. Moore's Law is the observation that the number of transistors in an integrated circuit should double every two years, which is something that did indeed happen for a few decades. But as more and more transistors are packed into the same area, electrons will start tunneling across different regions of the semiconductor and weaken or disrupt the fidelity of the electric currents that are necessary for logical computation. Chipmakers and tech companies are finding ways around these quantum limitations. One method they've used is simply making computer chips bigger; that's the strategy NVIDIA adopted with its latest generation AI chip, called Blackwell, which made its public debut in 2024 and is expected to pack an astonishing 208 billion transistors in a single unit, making it by far the world's most powerful computer chip as of this writing. But the broader strategy has been to rely on accelerated and parallel computing, which features the distribution of computing tasks across multiple cores packed into the same mainframe. By spreading out intensive computing operations across different computer chips working together, it becomes possible to generate ultra-fast computational speeds even with the constraints holding back Moore's Law.
18. Peter Atkins, *Four Laws That Drive the Universe* (Oxford: Oxford University Press, 2007), 51–52. The Carnot limit, or Carnot efficiency, is the highest possible efficiency that an ideal thermodynamic engine can have. As real-life engines are by definition not ideal, they will never be able to surpass the efficiency barrier of the Carnot limit.
19. John Bellamy Foster, *Ecology Against Capitalism* (New York: Monthly Review Press, 2002), 94.
20. Ibid.
21. For a prominent paper, see Kenneth Gillingham, David Rapson, and Gernot Wagner, "The Rebound Effect and Energy Efficiency Policy" *Review of Environmental Economics and Policy* 10 (2016). Also refer to Terry Barker et al., "The macroeconomic rebound effect and the world economy," *Energy Efficiency* 2 (2009): 411–27.

22. Carey King, *The Economic Superorganism* (New York: Springer International, 2020), 230.
23. Steven Sorrell, *The Rebound Effect* (London: UK Energy Research Centre, 2007), 92.
24. Fiona Harvey, "UN Warns of 'Unacceptable' Greenhouse Gas Emissions Gap," *The Guardian*, October 31, 2017, https://www.theguardian.com/us.
25. Cathy Bussewitz, "Carbon dioxide emissions reached a record high in 2022," Associated Press, March 2, 2023.
26. Nijavalli H. Ravindranath and Jayant A. Sathaye, *Climate Change and Developing Countries* (New York: Springer, 2006), 35.
27. Zhen-He He et al., "ATP Consumption and Efficiency of Human Single Muscle Fibers with Different Myosin Isoform Composition," *Biophysical Journal* 79 (2000): 945–61.
28. On the efficiency of internal combustion engines, see Efstathios E. Stathis Michaelides, *Alternative Energy Sources* (New York: Springer, 2012), 411. For coal-fired power plants, see R. Sandström, "Creep Strength of Austenitic Stainless Steels for Boiler Applications," in *Coal Power Plant Materials and Life Assessment*, ed. A. Shibli (Amsterdam: Elsevier, 2014), 128. On the efficiency of photovoltaic cells, see Friedrich Sick and Thomas Erge, *Photovoltaics in Buildings* (London: Earthscan, 1996), 14.
29. Robert T. Balmer, *Modern Engineering Thermodynamics* (Cambridge: Academic Press, 2011), 454.
30. Will Oremus, "How Green Is a Tesla, Really?," *Slate*, September 9, 2013, http://slate.com.
31. For an excellent discussion of these issues, see Carey King, *The Economic Superorganism* (New York: Springer International, 2020), 59–115.
32. Ibid., 96–99.
33. Liz Alderman, "French Nuclear Power Crisis Frustrates Europe's Push to Quit Russian Energy," *New York Times*, June 19, 2022.
34. Amory Lovins and M. V. Ramana, "Three Myths About Renewable Energy and the Grid, Debunked," *Yale Environment 360* (2021). https://e360.yale.edu/.
35. Liz Alderman, "French Nuclear Power Crisis Frustrates Europe's Push to Quit Russian Energy," *New York Times*, June 19, 2022.
36. King, *The Economic Superorganism*, 93.
37. Ibid.
38. Ibid.
39. See Benjamin K. Sovacool, "Valuing the greenhouse gas emissions from nuclear power: A critical survey," *Energy Policy* 36 (2008): 2950–63.

Also refer to Ethan Warner and Garvin Heath, "Life Cycle Greenhouse Gas Emissions of Nuclear Electricity Generation," *Journal of Industrial Ecology* 16 (2012): S73–S92.
40. For a recent comprehensive study on the subject, see Francesco Pomponi and Jim Hart, "The greenhouse gas emissions of nuclear energy," *Applied Energy* 290 (2021): 116743.
41. John Bellamy Foster, "Making War on the Planet," *Monthly Review* 70 (2018): 1–10.
42. Ibid.
43. IEA, *Bioenergy with Carbon Capture and Storage* (Paris: IEA, 2022), https://www.iea.org/.
44. Foster, "Making War on the Planet," 1–10.
45. Vaclav Smil, "Global Energy: The Latest Infatuations" *American Scientist* 99 (2011).
46. Mafalda Silva and Ingunn Saur Modahl, "The inventory and life cycle data for Norwegian hydroelectricity," Norwegian Institute for Sustainability Research, 2019, https://norsus.no/.
47. Mira Rojanasakul and Max Bearak, "Is It a Lake, or a Battery? A New Kind of Hydropower Is Spreading Fast," *New York Times*, May 2, 2023.
48. See Rachel DuRose, "The terrible paradox of air pollution and climate change," Vox, September 17, 2023. https://www.vox.com/.
49. Dan Gearino, "A New Battery Intended to Power Passenger Airplanes and EVs, Explained," *Inside Climate News*, May 18, 2023, https://insideclimatenews.org/.

8. Energy and Technology in the Social Sphere

1. Karl Marx and Friedrich Engels, *The Communist Manifesto* (New York: Penguin, 1998), 50.
2. Stephanie Kelton, *The Deficit Myth* (New York: Hachette Book Group, 2020), 25-26.
3. For example, see David Dayen, *Monopolized: Life in the Age of Corporate Power* (New York: New Press, 2020).
4. See Tim Jackson, "The Post-Growth Challenge: Secular Stagnation, Inequality, and the Limits to Growth," *Ecological Economics* 156 (2019): 236–46.
5. Thomas O. Wiedmann et al., "The Material Footprint of Nations," *Proceedings of the National Academy of Sciences* 112 (2013): 6271–76.
6. Daron Acemoglu and Pascal Restrepo, "Automation and New Tasks: How Technology Displaces and Reinstates Labor," *Journal of Economic Perspectives* 33 (2019): 3–30.
7. Robert Walton, "US electricity load growth forecast jumps 81% led by

data centers, industry," *Utility Dive*, December 13, 2013. https://www.utilitydive.com/.
8. Ashley Nunes, "Automation Doesn't Just Create or Destroy Jobs—It Transforms Them," *Harvard Business Review*, November 2, 2021, https://hbr.org/.
9. Avi Asher-Schapiro, "Uber's Business Model: Screwing its Workers," *In These Times*, September 26, 2014.
10. David Gerard and Lester Lave, "Implementing technology-forcing policies: The 1970 Clean Air Act Amendments and the introduction of advanced automotive emissions controls in the United States," *Technological Forecasting and Social Change* 72 (2005): 761–78.
11. Jeremy Deaton, "The Stunning Hypocrisy of U.S. Automakers," Nexus Media News, 2018, https://nexusmedianews.com/.
12. See Markus Krajewski, "The Great Lightbulb Conspiracy," *IEEE Spectrum*, September 24, 2014, https://spectrum.ieee.org/.
13. See Bobby Allyn, "Apple Agrees to Pay $113 Million to Settle 'Batterygate' Case Over iPhone Slowdowns," NPR, November 18, 2020, https://www.npr.org/.
14. Julia Alexander, "Disney is ending its vault program, giving Disney+ a huge boost in the streaming wars," *The Verge*, March 7, 2019, https://www.theverge.com/.
15. Julie Creswell, "Your Steak Is More Expensive, but Cattle Ranchers Are Missing Out," *New York Times*, June 23, 2021.
16. Cade Metz and Karen Weise, "Microsoft to Invest $10 Billion in OpenAI, the Creator of ChatGPT," *New York Times*, January 23, 2023.
17. Hannah Olivennes, "Toblerone alters shape of 2 Chocolate Bars, and Fans Are Outraged," *New York Times*, November 8, 2016.
18. See Espen Moen, Fredrik Wolfsberg, and Øyvind Aas, "Price dispersion and the role of stores," *Scandinavian Journal of Economics* 122 (2020): 1181–1206.
19. Ibid.
20. Seth Hanlon, "Accelerated Depreciation," Center for American Progress, March 23, 2011, https://www.americanprogress.org/.
21. Steve Wamhoff and Matthew Gardner, "Twenty-Three Corporations Saved $50 Billion So Far Under Trump Tax Law's Bonus Depreciation that Many Lawmakers Want to Extend," Institute on Taxation and Economic Policy, November 2022, https://itep.org/.
22. For an excellent overview of Amazon's monopolistic practices, see Lina M. Kahn, "Amazon's Antitrust Paradox," *Yale Law Journal* 126 (2017): 710–805.
23. Joseph Pisani, "Despite green pledges, Amazon's carbon footprint grew," Associated Press, June 23, 2020, https://apnews.com/.

24. David Croteau, William Hoynes, and Clayton Childress, *Media/Society: Technology, Industries, Content, and Users* (Thousand Oaks, CA: SAGE Publications, 2021), 40.
25. Ibid.
26. Alan Devlin, *Fundamental Principles of Law and Economics* (New York: Routledge, 2014), 386.
27. See Stephen Labaton, "Five Music Companies Settle Federal Case on CD Price-Fixing," *New York Times*, May 11, 2000.
28. See Dennis Carlton et al., "The Economics of the LCD Cartel: Organization, Incentives and Practical Challenges," eds., Joseph E. Harrington and Maarten Pieter Schinkel, *Cartels Diagnosed: New Insight on Collusion* (Cambridge, Cambridge University Press, 2024).
29. See John M. Connor, "Twilight Prosecutions of the Global-Auto Parts Cartels," American Antitrust Institute, 2019.
30. Peter Grier, "In the Beginning, There Was ARPANET," *Air Force Magazine*, January 1, 1997, https://www.airandspaceforces.com/.
31. Chris Rizos, "Introducing the Global Positioning System," in *Manual of Geospatial Science and Technology*, ed. John D. Bossler, John R. Jensen, and Robert B. McMaster (London: Taylor & Francis, 2022), 77.
32. Adi Robertson, "Tesla repays $465 million government green energy loan ahead of schedule," *The Verge*, May 22, 2013, https://www.theverge.com/.
33. Kenneth Flamm, *Creating the Computer: Government, Industry, and High Technology* (Washington, DC: Brookings Institution, 1988), 16.
34. Ibid.
35. Charles Wessner, *Securing the Future: Regional and National Programs to Support the Semiconductor Industry* (Washington, DC: National Academies Press, 2003), 13.
36. Lisa Wang, "Government to sell TSMC shares," *Taipei Times*, November 21, 2006, https://www.taipeitimes.com/.
37. See Mariana Mazzucato, *The Entrepreneurial State: Debunking Public vs. Private Sector Myths* (London: Anthem Press, 2013).
38. Vaclav Smil, *Energy and Civilization* (Cambridge, MA: MIT Press, 2017), 395.
39. Ibid.
40. Benjamin Sovacool, "How Long Will It Take? Conceptualizing the temporal dynamics of energy transitions," *Energy Research & Social Science* 13 (2016): 202–15.
41. Ibid.

9. The Industrialization of Britain: A Case Study

1. See Jeremy Rifkin, *The Third Industrial Revolution* (New York: Palgrave Macmillan, 2011).

2. David S. Landes, *The Unbound Prometheus* (Cambridge: Cambridge University Press, 1969), 6.
3. See Robert C. Allen, *The British Industrial Revolution in Global Perspective* (Cambridge: Cambridge University Press, 2009), 13–14.
4. Ibid., 125.
5. Ibid.
6. Ibid., 125–26.
7. Ibid., 130.
8. Ibid., 34.
9. Ibid.
10. James Belich, *The World the Plague Made* (Princeton: Princeton University Press, 2022), 123–24.
11. Ibid.
12. Ibid., 124.
13. Ibid., 125.
14. Ibid.
15. Ibid.
16. Ibid., 125–126.
17. Ibid., 125–126.
18. Ibid., 125–126.
19. Allen, *The British Industrial Revolution in Global Perspective*, 61–63.
20. Stewart Ross, *The Industrial Revolution* (London: Evans Brothers Limited, 2008), 16.
21. T. F. Walsh, *America 2100: After Fossil Carbon* (Raleigh: Lulu Publishing, 2015), 68.
22. Allen, *The British Industrial Revolution in Global Perspective*, 109–10.
23. See John U. Nef, "An Early Energy Crisis and Its Consequences," *Scientific American* 237 (1977): 140–51.
24. Brinley Thomas, *The Industrial Revolution and the Atlantic Economy: Selected Essays* (New York: Routledge, 1993), 16.
25. See George Selgin and John L. Turner, "Strong Steam, Weak Patents, or the Myth of Watt's Innovation-Blocking Monopoly, Exploded," *Journal of Law and Economics* 54 (2011): 841–61.
26. See John H. Lienhard, "The Engines of Our Ingenuity, University of Houston," February 18, 1999, https://engines.egr.uh.edu/talks/powersir.
27. See H. W. Dickinson and Arthur Titley, *Richard Trevithick: The Engineer and the Man* (Cambridge: Cambridge University Press, 2010), 43–60.
28. Allen, *The British Industrial Revolution in Global Perspective*, 173.
29. Bruce Hunt, *Pursuing Power and Light* (Baltimore: Johns Hopkins University Press, 2010), 10.
30. For data on efficiencies and power outputs, see Richard Brown, *Society and Economy in Modern Britain 1700-1850* (Milton Park: Taylor &

Francis, 2002), 60. Also refer to Robert Balmer, *Modern Engineering in Thermodynamics* (Burlington: Academic Press, 2011), 454, 466. For data on watermills, see Terry Reynolds, *Stronger Than a Hundred Men: A History of the Vertical Water Wheel* (Baltimore; Johns Hopkins University Press, 1983), 123.
31. Ronald Findlay and Kevin O'Rourke, *Power and Plenty* (Princeton: Princeton University Press, 2007), 320.
32. See Paul Bairoch, *Economics and World History: Myths and Paradoxes* (Chicago: The University of Chicago Press, 1993).
33. See Joseph Inikori, *Africans and the Industrial Revolution in England: A Study in International Trade and Economic Development* (Cambridge: Cambridge University Press, 2002).
34. Jim Harter, *World Railways of the Nineteenth Century* (Baltimore: Johns Hopkins University Press, 2005), 2.
35. Andrew Herod, *Geographies of Globalization: A Critical Introduction* (West Sussex: Wiley-Blackwell, 2009), 117.
36. See Utsa Patnaik and Prabhat Patnaik, "The Drain of Wealth," *Monthly Review* 72 (2021).
37. Brinley Thomas, *The Industrial Revolution and the Atlantic Economy: Selected Essays* (New York: Routledge, 1993), 14.
38. Ibid., 14-15.
39. Ibid.
40. Lee T. Wyatt III, *The Industrial Revolution* (Westport, CT: Greenwood Press, 2009), 53.
41. Anthony Page, *Britain and the Seventy Years War* (London: Palgrave Macmillan, 2017), 95.
42. Ibid.
43. Ibid.
44. Jaume Ventura and Hans-Joachim Voth, "Debt into Growth: How Sovereign Debt Accelerated the First Industrial Revolution," National Bureau of Economic Research (2015), 2, https://www.nber.org/.
45. Martin Hutchinson and Kevin Dowd, "The Apotheosis of the Rentier: How Napoleonic Wars Finance Kick-Started the Industrial Revolution," *Cato Journal* 38 (2018): 255–78.
46. Ventura and Voth, "Debt into Growth: How Sovereign Debt Accelerated the First Industrial Revolution," 2.
47. It's often mistranslated as "*primitive* accumulation" in English.
48. See Amitav Ghosh, *The Nutmeg's Curse* (Chicago: University of Chicago Press, 2021), 1–25.

10. A New Vision

1. See Richard Lee, *The Dobe Ju/'hoansi* (Boston: Cengage Learning, 1984).

2. See E. P. Thompson, "Time, Work-Discipline, and Industrial Capitalism," *Past & Present* 38 (1967): 56–97.
3. Barbara Rogoff, *The Cultural Nature of Human Development* (New York: Oxford University Press, 2003), 4.
4. Ibid., 5.
5. Ibid.
6. Ibid.
7. Ibid.
8. Adrianna Rodriguez, "Americans are lonely and it's killing them," *USA Today*, December 24, 2023.
9. Marcel Mazoyer and Laurence Roudart, *A History of World Agriculture* (New York: Monthly Review Press, 2006), 30.
10. Mathis Wackernagel and William Rees, *Our Ecological Footprint* (Gabriola, BC: New Society Publishers, 1996), 11.
11. The figures for the United States and all other countries in this chapter come from BP, *Statistical Review of World Energy 2022*, 8. Note that BP gives its primary energy consumption figures in exajoules, so those have to be converted to kilocalories to make them compatible with the numbers I'm using here.
12. Ibid., 13.
13. Ibid.
14. A notable essay on the subject is from Michael A. Lebowitz, "Proposing a Path to Socialism," *Monthly Review* 65 (2014).
15. For white collar crime figures in general, see David Friedrichs, *Trusted Criminals: White Collar Crime in Contemporary Society* (Boston, Cengage Learning, 2009). For tax evasion numbers, see William Gale and Aaron Krupkin, "How big is the problem of tax evasion?" Brookings Institution, April 9, 2019, https://www.brookings.edu/. The authors estimate that tax evasion in the United States in 2018 may have totaled roughly $600 billion.
16. See Lu Aye et al., "Life cycle greenhouse gas emissions and energy analysis of prefabricated reusable building modules," *Energy and Buildings* 47 (2012): 159–68.
17. For just one example, see Hans Wiesmeth, *Implementing the Circular Economy for Sustainable Development* (Amsterdam: Elsevier, 2021).
18. See Greenpeace, "Circular Claims Fall Flat Again," October 2022, https://www.greenpeace.org/usa/.
19. See Intan Suwandi, *Value Chains: The New Economic Imperialism* (New York: Monthly Review Press, 2019).
20. David Dollar, "Value chains transform manufacturing—and distort the globalization debate," International Monetary Fund, June 2019, https://www.imf.org/en/Home.

21. See John Bellamy Foster, Robert W. McChesney, and R. Jamil Jonna, "The Global Reserve Army of Labor and the New Imperialism," *Monthly Review* 63 (2011).
22. Simon Bowers, "Apple's cash mountain, how it avoids tax, and the Irish link," *The Irish Times*, November 6, 2017, https://www.irishtimes.com/.
23. Ibid.
24. For example, see Thin Lei Win, "Multinational companies account for nearly a fifth of global CO_2 emissions, researchers say," Reuters, September 8, 2020, https://www.reuters.com/.

11. The Ecological State

1. Steven Clifford, *The CEO Pay Machine* (New York: Blue Rider Press, 2017), 34.
2. Sarah Anderson, "The failure of Bill Clinton's CEO pay reform," *Politico*, August 31, 2016. https://www.politico.com/.
3. See Heather Long, "The Federal Reserve Has Pumped $2.3 Trillion into the U.S. Economy. It's Just Getting Started," *Washington Post*, April 29, 2020.
4. Michael Birnbaum, "Coronavirus Hits European Economies but Governments Help Shield Workers," *Washington Post*, April 30, 2020.
5. This is often known as *Pareto efficiency*, after the Italian economist Vilfredo Pareto.
6. See Ludwig von Mises, *Economic Calculation in the Socialist Commonwealth* (Auburn, AL: Ludwig von Mises Institute, 2014).
7. For a concise description of the Inca imperial economy, see Gordon Francis McEwan, *The Incas: New Perspectives* (New York: W. W. Norton, 2008), 87–88.
8. See Leigh Phillips, *The People's Republic of Walmart* (New York: Verso, 2019).
9. See Jonathan Nitzan and Shimshon Bichler, *Capital as Power* (Abingdon: Routledge, 2009).
10. Stephen Martin and David Parker, "Privatization and Economic Performance throughout the UK Business Cycle," *Managerial and Decision Economics* 16 (1995): 225–37.
11. Arunava Bhattacharyya, C. A. K. Lovell, and Pankaj Sahay, "The Impact of Liberalization on the Productive Efficiency of Indian Commercial Banks," *European Journal of Operational Research* 98 (1997): 332–45.
12. For example, see Sergei Guriev, Anton Kolotilin, and Konstantin Sonin, "Determinants of Nationalization in the Oil Sector: A Theory and Evidence from Panel Data," *Journal of Law, Economics, and Organization* 27 (2011): 301–23.

13. See Matt Huber and Fred Stafford, "In Defense of the Tennessee Valley Authority," *Jacobin*, April 4, 2022, https://jacobin.com/.
14. See ISLR, "Public Banks: Bank of North Dakota," Institute for Local Self-Reliance, 2023, https://ilsr.org/.
15. Edward Henniker-Major, "Nationalization: The Anglo-Iranian Oil Company," *Moral Cents: The Journal of Ethics in Finance* 2 (2013).
16. Philip Barker, *Religious Nationalism in Modern Europe* (New York: Routledge, 2008), 9.
17. Ibid.
18. R. R. Palmer and Joel Colton, *A History of the Modern World* (New York: McGraw Hill, 1995), 148–49.
19. John Keane, *Tom Paine: A Political Life* (New York: Grove Press, 1995), 446.
20. The great scholar István Mészáros wrote many works on the transition to a post-capitalist society. For a notable reference, see István Mészáros, *Beyond Capital: Toward a Theory of Transition* (New York: Monthly Review Press, 2000).
21. For just one example, see "Liberal Democracy" from the European Center for Populism Studies. https://www.populismstudies.org/
22. In Book 3 of *Politics*, Aristotle writes: "Tyranny…is monarchy exerting despotic power over the political community; oligarchy is when the control of the government is in the hands of those that own the properties; democracy is when on the contrary it is in the hands of those that do not possess much property, but are poor." Refer to the translation in the online edition of the Perseus Digital Library from Tufts University, http://www.perseus.tufts.edu/hopper/.
23. See Thomas Ferguson, *Golden Rule: The Investment Theory of Party Competition and the Logic of Money-Driven Political Systems* (Chicago: University of Chicago Press, 1995).
24. See John Burnheim, *The Demarchy Manifesto* (Exeter: Andrews UK Limited, 2016).
25. See Ken Burns, "War Production" from the documentary *The War*, PBS, 2007, https://www.pbs.org/.
26. David J. Siemers, *The Myth of Coequal Branches* (Columbia: University of Missouri Press, 2018), 48.

12. The Valerist Cosmopole

1. Garrett Hardin, "The Tragedy of the Commons," *Science* 162 (1968): 1243–48.
2. Brett Frischmann, Alan Marciano, and Giovanni Battista Ramello, "Tragedy of the Commons after 50 Years," *Journal of Economic Perspectives* 33 (2019): 211–28.

3. Ian Angus, *The War Against the Commons* (New York: Monthly Review Press, 2023).
4. See Elinor Ostrom, *Governing the Commons: The Evolution of Institutions for Collective Action* (Cambridge: Cambridge University Press, 1990).
5. See Benjamin Neimark, Oliver Belcher, and Patrick Bigger, "The US military is a bigger polluter than more than 100 countries combined," *Quartz*, June 28, 2019, https://qz.com/.
6. See Johannes Friedrich et al., "This Interactive Chart Shows Changes in the World's Top 10 Emitters," World Resources Institute, March 2, 2023, https://www.wri.org/.
7. Felix Chang, "China's Rare Earth Metals Consolidation and Market Power," Foreign Policy Research Institute, March 2, 2022, https://www.fpri.org/.
8. See Vladimir Lenin, *Imperialism: The Highest Stage of Capitalism* (London: Penguin Books, 2010). Also see Wallerstein's four volumes on the development of the modern global economy, *The Modern World-System*.
9. See the first volume of Wallerstein's four-volume masterpiece on the modern world system: Immanuel Wallerstein, *The Modern World-System I* (Berkeley: University of California Press, 2011).
10. See Susanne Soederberg, *The Politics of the New International Financial Architecture* (London: Zed Books, 2004), 33. Also refer to Joseph Stiglitz, *Globalization and Its Discontents* (New York: W. W. Norton, 2003).
11. See Manmohan Agarwal and Dipankar Sengupta, "Structural Adjustment in Latin America," *Economic and Political Weekly* 34 (1999): 3129–36.

Index

absolute decoupling, 116
accelerated depreciation for taxes, 260–61
Acemoglu, Daron, 137–43, 253
adenosine triphosphates, 20
Advanced Research Projects Agency (ARPA), 265
aerobic organisms, 147
afforestation, 239–40
Afghanistan, 174
aggregate flow of energy, 192–95
aggregate flow rates (AFR), 193–94, 214; after Second World War, 271
aggregate production functions, 91
aggregation problem, 46–47
agriculture: automation in, 255; development of, 14–15, 27; energy use by, 171; in GDP, 106; during Industrial Revolution, 282; Marx on, 198; in Netherlands, 9–10
air pollution, deaths caused by, 30, 158–59
Aka (people), 308
Allen, Robert, 275, 284
Amazon (firm), 119, 261–62

Amsterdam Stock Exchange, 301
anaerobic organisms, 147
Ancient Greece, 135–36
Anglo-Iranian Oil Company (AIOC), 329–30
An Lushan Rebellion (China), 165–67
Anthropocene era, 173
anthropomass, 171
Apple (firm), 119, 258, 317–18
Aramco (firm), 328
Argentina, 151, 174
Aristotle, 334
Arrow, Kenneth, 84
artificial intelligence (AI), 150, 254
asset bubbles, 47–48
AT&T (firm), 262, 267
Australia, 303
autocatalytic sets, 220
automation, 253–56
automobiles: auto parts cartel, 263; catalytic converters for, 256–57; as contingent technology, 228
autopoiesis, 220
autotrophs, 20
Awami League (East Pakistan), 11

backfire (rebound effect), 231–32
Bailey, Ronald, 160–63
Bairoch, Paul, 295
Banda (people), 301
Bangladesh (East Pakistan), 11
banking, 314
Belich, James, 278
Bell Labs (firm), 262, 265–67
Bengal Famine (India), 167
Biden, Joe, 148
biodiversity, 153–54
bioenergy carbon capture and storage (BECCS), 240–41
biofuels, 121
biology, entropic imperative in, 24
biomass, 171–72
bionomic disruptions, 30
biosphere, humans and, 170–75
Black Death, 141, 221, 277–80
Böhm-Bawerk, Eugen von, 94–98, 101
boilers, 218
Bologna (Italy), 279
Boulton, Matthew, 287
Brazil, 243
British Empire, 161
Brunk, Gregory, 68
Bureau of Economic Analysis (BEA), 53, 58–59
Burnheim, John, 335
Byzantine Empire, 140

capital, productivity of, 90–91
capitalism: artificial intelligence in, 254; central planning under, 327; childcare under, 307–8; creative destruction as force in, 137; degeneration of natural world required for, 198; ecological costs of, 148–49; emergence of, 27; energy consumption in, 28; growth essential to, 103; metabolic rift theory on, 198–99; monopolies and cartels in, 261–63; spectralization of technology under, 228–29; violence in development of, 302; warfare and, 163; waste created by, 316
capital stock, 91–93
carbon capture and storage (CCS), 240, 241
carbon dioxide: in biomass, 172; carbon dioxide removal (CDR), 239–40; emissions of, 119, 233, 262
carbon emissions, 126
carbon offsets, 120
carbon taxes, 269
carnivores, 20
Carnot, Sadi, 178
Carnot cycle, 230
Carnot engines, 178
Carson, Rachel, 199
cartels, 262–63
catalytic converters, 256–57
CATL (firm), 244
causation, 79–81
Central Intelligence Agency (CIA), 330
chain-weighting, 53–58
chickens, 154
childcare, 306–8
Chile, 151
China, 60, 163; afforestation in, 240; An Lushan Rebellion in, 165–67; droughts in, 174; economic growth in, 149; government-owned banks in, 328; life expectancy in, 62; rivalry between United States and, 343–44; during Song dynasty, 206–9
chlorofluorocarbons (CFCs), 29–30
circular economy, 315
Citigroup (firm), 99
civic nationalism, 331
Clark, Brett, 198–99
Clark, John Bates, 87–88
classical economics, 70

climate, 14; integrated assessment models on, 106–7; tipping points in, 107–8
climate change, *see* global warming
Clinton administration, 323
coal: converted into electricity, 184; efficiency of, 181; energy from burning of, 283; Jevons paradox in consumption of, 231; as primary energy supply, 267; steam engines used in mining of, 285–92; sulfur dioxide emissions from, 160; used in China, 207, 208; used in England, 209, 275–76, 281
cocoa beans, 345
Coen, Jan, 163
coke (fuel), 297
consols (consolidated annuities), 298, 299
Consumer Price Index (CPI), 47
consumption of energy, 186–92
contingents, 228
conversional networks (coronets), 181–86
Cort, Henry, 297
Cosmopole, 347–52
cotton industry, 292–93
covid (coronavirus) pandemic, 100, 201; childcare during, 307; economic policies during, 324; vaccines for, 266
Cromwell,, Oliver, 297
Cuba, 36, 62
cultural hegemony, 135
cultural school, 126–36
currencies, 250; of China under Song dynasty, 206; exchange rates for, 59
cyanobacteria, 146
Cyclone Bhola (1970), 11

da Gama, Vasco, 300
Dai, Aiguo, 158

Darby, Abraham, 297
deaths, *see* mortality
De Beers (firm), 80, 82
Debreu, Gérard, 83–84
decoupling process, 115–26
deforestation, 172, 207, 209–11; in England, 284–85
demand, supply and, 78–87, 201
demarcation problem, 81
demarchy, 335
democracies, 164, 334–35
demopoles (democratic polis), 338
Diamond, Jared, 143, 144, 212
diamonds, 80
diffusion, 223
discount rates, 108
diseases, deaths from, 11–12
Disney (firm), 258–59
dissipation, 21–25
distribution, 87–93
droughts, 158, 174
Duke Energy (firm), 119
Dutch East India Company (VOC), 300–302
Dynamic Integrated Climate-Economy model (DICE), 106, 107

Earth, changes to environment caused by humans, 28–29
Earth System, 28
Easter Island, 205–6
East Pakistan (Bangladesh), 11
ecodynamic synthesis, 16–17, 196–97, 220; Industrial Revolution and, 272–82
ecological economics, 64
ecological footprints, 310–13
ecological society, 321
ecological state, 321–22
ecological valence, 309–10
ecology, 197; disruptions of, in human history, 30; humanity and, 200–204
economic growth, 26–27, 33–38,

INDEX 387

103–4; aggregate flow rates as measurement of, 193–94; cultural explanations of, 126–36; decoupling from human needs, 115–26; energy consumption correlated with, 124; future of, 303–4; institutional school on, 136–45; neoclassical economic theory on, 104–9; scale and efficiency in, 63–69; technological change and, 214–17, 220–29; in valerist economies, 311
economics: neoclassical theory in, 70–74; physics and, 26–29; units in, 41
economies: aggregate flow of energy in, 192–95; circular, 315; as conversional network, 182; natural world tied to, 196–97
ecosphere, waste and energy absorbed by, 221
Efe (people), 308
efficiency: energy and, 177–81; scale and, 65–67; technological innovation and, 229–34
Ehrlich, Paul, 149
elections, 334
electricity: consumed by artificial intelligence, 254; efficiency of, 180–81; fossil fuels converted into, 183, 184; measuring production of, 187
electric vehicles (EVs), 244
emergence, 73
employment, 318
energy, 18–20; aggregate flow of, 192–95; from burning coal, 283; consumed by economies, 26; conversional network for, 181–86; conversion and consumption of, 186–92; conversions of, 176; economic growth and, 105; efficiency and, 177–81; essential for economic activity, 121–23; German policies on, 110–11; Jevons paradox in efficiency of, 230–32; per capita consumption of, 27; productivity and, 64–67; solar, 113; spectralization of, 222–29; transitions of, 265–70; units of, 25
energy-complexity spiral, 67
energy flows, 19–20
energy quality, 192
energy return on investment (EROI), 189
Engels, Friedrich, 137, 249
England, Jeremy, 24–25
England (country): deforestation of, 209; Industrial Revolution starting in, 131–32, 144, 222, 275–81; Magna Carta in, 142; spectralization of economy of, 282–95; *see also* Great Britain
English Civil War, 142
Enlightenment philosophy, 129
entropic imperative, 24
entropy, 21–25
environment, human-caused changes in, 28–29
environmental determinism, 17
epidemics, 221
ethnic nationalism, 331
eutrophication, 10
exchange rates, 59
exergy, 178–80, 290
extinctions, 152–53; caused by human habitation, 303; *see also* mass extinction events

family, under capitalism, 307
fascism, 162, 331
Federal Reserve, chain-weighting GDP by, 55–58
Feigenbaum, Susan, 82
fertilizers, 10
finance system, 314–15
First Industrial Revolution, 271
Fisher, Irving, 72, 94

Fisher index, 54, 55
Fix, Blair, 53
flywheels, 219
Ford Motors (firm), 222–23
Fore (people), 308
forests, 172; afforestation of, 239–40; deforestation of, 207, 209–11
fossil fuels: converted into electricity, 183, 184; in Industrial Revolution, 282–83; mortality and, 157–60; pollution generated by burning of, 243–44; power densities of, 191
Foster, John Bellamy, 198–99
Fourth Crusade, 140
Fox, Shadrach, 297
France, 37; heat waves in, 174; in Hundred Years' War, 142; nationalism in, 331–32; nuclear power in, 237, 268; property rights in, 275
free markets, 324
French Revolution, 331–32
friendships, 308

gas prices, 80
general equilibrium theory, 83–84
geoengineering, 235, 239–43
Georgescu-Roegen, Nicholas, 199
Germany, 110–11, 238
global labor arbitrage, 317
Global North, 345
Global Positioning System (GPS), 265
Global South, 345
global value chains, 317
global warming, 175; caused by hydrofluorocarbons, 30; costs of, 154; mortality linked to, 155–59
Glorious Revolution (England), 140, 275
governments: businesses owned by, 328; energy transitions managed by, 269–70; impacts on markets by, 324; parliamentary form of, 274–75; research and development funded by, 265–67; structure of valerist state, 330–40
Gramsci, Antonio, 135
Great Britain, 37; economic growth in, 138–40; in Hundred Years' War, 142–43; Industrial Revolution starting in, 131–32, 144; Iranian government overthrown by United States and, 330; Magna Carta in, 142; in Napoleonic Wars, 296; national debt of, 298–99; trade during Second Industrial Revolution in, 296–98; *see also* England
Great Recession, 257
Greece (ancient), 135–36, 334
greenhouse gas (GHG) emissions, 30, 117–20, 125–26, 233; from China, 343; geoengineering strategies for control of, 239–43; from United States, 123–24; *see also* carbon dioxide
Gross Domestic Product (GDP), 34–38; agriculture in, 106; decoupling and, 116; measurement of, 41; varieties of, 52–63
Grubler, Arnulf, 191
Guisso, Richard, 166
Guo, Yuming, 157
Guterres, Antonio, 175

Hall, Charles, 189
Hansen, James, 241
Hansen, Valerie, 166
Hardin, Garrett, 341–42
Hawaii, 144
herbivores, 20
heterotrophs, 20
Holocene era, 173
Honshu (Japan), 210
House of Representatives (U.S.), 337
Hughes, Thomas, 217–18

humans: biosphere and, 170–75; changes to Earth's environment caused by, 28–29; consumption of energy by, 25; ecodynamic synthesis of nature with, 16–17; ecological disruption in history of, 30; ecology and, 200–204; impacts of nature on, 10–13; prospects for, 147–52
Hundred Years' War, 142–43, 168
Hussite Wars, 168
hydroelectric energy, 242–43; *see also* renewable energy sources
hydrofluorocarbons (HFCs), 30

ideogenesis, 133
imperial dynamics theory, 169
imperialism, 296, 344–45
India: Bengal Famine in, 167; consumption rate in, 311–12; in Cosmopole, 348; textile industry in, 298
Indonesia, 12, 121, 301
industrialization, 271–72, 296
Industrial Revolution, 127, 131–32; ecodynamic synthesis and, 272–82; energy sources used in, 191; in England, 138–40, 143, 222; rapid rate of growth during, 304; steam as dominant energy source in, 292–95
inflation, 43–44; calculating, 48–52; measurement of, 54–55
Inikori, Joseph, 295
insects, 153
institutional school, 136–45
integrated assessment models (IAMs), 106
Intel (firm), 47
interdependence, in supply and demand, 84–85
interest rates, 94, 96, 99
Intergovernmental Panel on Climate Change (IPCC), 106

International Geosphere-Biosphere Programme, 28
International Monetary Fund (IMF), 35, 345–46
International Union for the Conservation of Nature (IUCN)., 152, 153
Internet, 265–70
invisible hand theory, 81
Iran, 329–30
Iraq, 161
iron: in China under Song dynasty, 206–7; production of after Black Death, 279; smelting of, 297–98
Italy, 36–37, 313

Jakarta (Indonesia), 12
Japan, 144–45, 209–12
Jevons, William Stanley, 71, 230–31
Jevons paradox, 230–32
Jin dynasty (China), 208
John (king, England), 142
joules (units of energy), 25, 193
JPMorgan (firm), 120
judiciary, 337
Ju/'hoansi (people), 305–6
Jurchen (people), 208

Kaifeng (China), 208
Kant, Immanuel, 149
Kauffman, Stuart, 220
keyboards, 219
Keynes. John Maynard, 149–50
kilocalories (food calories; units of energy), 25
kinetic energy, 19
King, Carey, 66, 231
Kinney, Patrick, 156
Kuznets, Simon, 37
Kwara'ae (people), 307–8

Landes, David, 272–73
Laspeyres index, 54
League of Nations, 350

Lee, Richard, 305
legislatures, 336–37
Lelieveld, Jos, 159
Lenin, V. I., 344
liberal democracy, 334
liberalism, political economy of, 322–30
life expectancy, 61–62, 116
light bulbs, 257–58
liquid crystal display (LCD) panels, 263
Liverpool-Manchester railway line, 296
localization, 316–18
Lomborg, Bjørn, 152, 154–59
London (England): Black Death mortality in, 141; growth of population of, 143, 283; impact of Uber in, 255; Industrial Revolution in, 280–81

Magna Carta (1215), 142
Malaysia, 118–19
Malthus, Thomas, 149
mammals, 154
Mandelbaum, Michael, 33
Mankiw, Gregory, 35
Mantel, Rolf, 83–84
marginalists, 71–72; on impulsive financial decisions, 76–77; on productivity, 87–88; on time preference theory, 93–94
marginality, in production and distribution, 87
market prices, 44–45
marriage, 274
Marx, Karl, 63–64; on class conflict, 249; on "dead labor," 90; on metabolic rift, 111; on national debts, 324; on natural world, 197–98; on primary accumulation, 300; Schumpeter on, 137; on technology and labor, 218
mass extinction events, 30, 152;

first, 147; Quaternary Megafauna Extinction, 170–71; sixth, 153
material footprints, 253
materialism, culture and, 134, 135
Maturana, Humberto, 219–20
Maya civilization, 150, 161
Mazoyer, Marcel, 309
McNaught, William, 290
meatpacking industry, 259
mechanical energy, 19; dissipation of, 21
mechanical work, 19, 193
Menger, Carl, 71, 93
mesopoles, 338–39
metabolic rift theory, 111, 198–99
meter (unit of length), 39
methane (natural gas), 118, 123, 242
Mexico City (Mexico), 12
Mirowski, Philip, 71
Mises, Ludwig von, 162, 326, 327
Mnuchin, Steve, 33
modal adaptation and conversional spectralization (MACS), 220–29
modernization, 272–73
modularization, 315–16
Mokyr, Joel, 127–36
mollusks, 152–53
Mondelez International (firm), 259–60
money, 250–51; *see also* currencies
Mongols (people), 209
Mongol wars, 168
monopolies, 261–63
Montreal Protocol (1987), 29
Moore's Law, 230
Morris, Ian, 212
mortality: caused by air pollution, 30; climate change and, 155–60; from warfare, 165–69
Mosaddegh, Mohammad, 330
Mosler, Warren, 250
music industry, 263
Musk, Elon, 259

nagana (disease), 11–12
Napoleonic Wars, 296
nationalism, 330–33
nationalization, 314, 327
nations: distinguished from states, 330; *see also* governments
natural gas (methane), 118, 123, 183–84; as primary energy supply, 268–69
nature: ecodynamic synthesis of humans with, 16–17; human economies tied to, 196; impact on humans of, 10–13; Marx on, 197–98; metabolic rift between society and, 111; society embedded in, 18
Nauru, 112–13
Nef, John, 284
neoclassical economic theory, 70–74, 322–23; on distribution, 87; on growth, 104–9; on productivity, 88–90; on rationality, 77; substitution in, 109–14; supply and demand in, 78–87
Netherlands, 9–10, 268–69; Dutch East India Company of, 300–302
Newcomen, Thomas, 277, 285–88
New York (New Amsterdam), 301
Nigeria, 52–53
Nike (firm), 252
nitrogen pollution, 9–10
nomadic communities, 305–6
Nordhaus, William, 106, 107
Norway, 242–43
nuclear power, 236–38, 268
Nusantara (Indonesia), 12
nutmeg, 300–302

ocean fertilization, 240
Odum, Howard, 200
oil: extraction of, 182–83; as primary energy supply, 267–68
oligarchies, 334
Organization for Economic Cooperation and Development (OECD), 35
Ostrom, Elinor, 342
oxygen, 146–47
ozone layer, 29

Paasche index, 54
Pacey, Arnold, 207
Pakistan, 174
palm oil, 121
partial-substitution method of measuring primary energy, 187
patents, 216
Patterson, Murray, 180–81
pesticides, 199
pharmaceutical industry, 266
Phoebus (light bulb cartel), 257–58
phosphorus, 10
photosynthesis, 146
photovoltaics, 113
physical-energy content method of measuring primary energy, 187
physics: economics and, 26–29; units in, 39
phytomass, 171
pig iron, 297–98
Pindyck, Robert, 108–9
Pinker, Steven, 164–68
planned obsolescence, 258
plants: biomass of, 170, 172; phytomass of, 171
political economy, 322–30
Polynesians, 204–6
Popular Mechanics (magazine), 150
population, 28
Portugal, 300
post-capitalist society, 318–20; general principles of, 309–10; localization in, 316–18; modularization in, 315–16; socialization in, 314–15; valerism in, 310–15
potential energy, 19
power (rate of energy use), 25
power density, 190, 191

Preston, Samuel, 61
Preston curve, 61–62
price indexes, 44, 53–54
prices, 327; as units of measurement, 41–46; utility and, 74–78
primary accumulation, 300
primary energy consumption, 186
primary forms of energy, 186–87
prime movers, 227–28
private property, 323
privatization, 327
problem of time, 45
processing sequences, 220
production, 87–93; global value chains for, 317; during Second World War, 335
productivity: energy and, 64–67; neoclassical economic theory on, 88–90; total factor productivity, 217
progress, 151
property rights, 275
Puerto Rico, 174
purchasing power, 59–60

Qian, Taotao, 158
Quaternary Megafauna Extinction, 170–71

railroads, 296
Rapa Nui (people), 205–6
rationality, 77
Reagan, Ronald, 323
rebound effect, 231–32
recycling, 316
Red List of Threatened Species, 152, 153
Rees, William, 199–200, 310–11
reference years, 53
regeneration, 310
relative decoupling, 116
renewable energy sources, 241–45; future of, 236; measuring energy from, 187; transition to, 226–27

Republic of Letters (RoL), 130–31
research and development (R&D), 216–17, 266
Restrepo, Pascual, 253
reverse salients, 218
Robinson, James, 137–43
Robinson, Joan, 77
Roche, Jason, 140
Roman Empire, 169; collapse of, 67; consumption rate in, 312; Pinker on, 168
Roser, Max, 168–69
Roudart, Laurence, 309
Russia, 59, 160

Samuelson, Paul, 75
San groups (peoples), 305–6
Schrödingern Erwin, 24
Schumpeter, Joseph, 128, 137
secondary forms of energy, 186
Second Industrial Revolution, 271, 280; Great Britain during, 296–98
Second World War, 165, 167; technological change tied to, 266, 267; War Production Board during, 335
self-organized criticality (SOC), 68
Senate (U.S.), 337
sheep, 280
shipbuilding industry, 280, 297
silk industry, 279
silver, 276
Simon, Herbert, 150
slavery, 301
smallpox, 12
SMD theorem, 84
Smeaton, John, 286
Smil, Vaclav: on carbon capture and storage, 241; on energy transitions, 267, 270; on limits to growth, 69; on power density of energy sources, 190, 191; on prime movers, 227; on resources extracted by humans, 171

Smith, Adam, 127
Snow, John W., 33
social class, technological change and, 248–49
socialization, 314–15, 325–30
society, 86–87; in ecodynamic synthesis, 196; ecological footprints of, 312–13; ecological society, 321; embedded in nature, 18; metabolic rift between nature and, 111
solar energy, 113; *see also* renewable energy sources
solar radiation management (SRM), 239
Solow, Robert, 43, 104–5
Solyndra (firm), 265
Song dynasty (China), 206–9
Sonnennschein, Hugo, 83–84
Sovacool, Benjamin, 268
Soviet Union, 160, 330
Spaain, 313
Spain, 300
spectralization, 222–29; of English economy, 282–95
spice trade, 300
Sraffa, Piero, 94
stabilization, 311–14
states: distinguished from nations, 330; *see also* governments
steam engines, 218, 231, 234; as dominant energy source, 292–95; invention of, in England, 276–77; used in coal mining, 285–92
Steindel, Charles, 58, 59
subjective theory of value, 71
sulfur dioxide emissions, 160
Sun, 20
supply: demand and, 78–87; limits on, 200–201
Sweden, 279–80

Tainter, Joseph, 67
taxes: accelerated depreciation for, 260–61; carbon taxes, 269; on wool exports, in England, 284
technological change: class-power dynamics of, 248–49; future direction of, 235–46; growth and, 220–29; theories of, 217–20; tied to government spending, 266–67
technological innovation, 216; aims of, 247; in auto industry, 257; automation as, 253–55; efficiency and, 229–34
technology: economic growth and, 214–17; energy use and, 31
telecommunications industry, 262
tertiary forms of energy, 186
Tesla motors, 234
textile industry, 280
Thatcher, Margaret, 86
the Federal Reserve System, 98–99
thermodynamics: entropy and dissipation in, 23–24; limits to substitution in, 110, 114
Third Industrial Revolution, 271
Thirty Years' War, 331
Tikopia, 204–5
time preference theory, 93–94, 99–100
tipping points, 111; in climate, 107–8
Tokugawa shogunate (Japan), 210–12
total factor productivity (TFP), 217
trade: during Industrial Revolution, 295–97; material footprint of, 253; shipbuilding industry for, 280; spice trade, 300
transistors, 266–67
transportation, during Industrial Revolution, 282
Trenberth, Kevin, 158
Trevithick, Richard, 288–90
Trump, Donald, 148, 257, 260–61
tsetse flies, 11
TSMC (firm), 266

turnpikes, 282

Uber (firm), 255
unitarian states, 336
United Kingdom, see Great Britain
United Nations, 346, 350
United States, 161; accelerated depreciation for taxes in, 260–61; consumption rate in, 311; division of power in, 336–37; exergy efficiency of, 178; greenhouse gas emissions in, 123–24; hegemonic power of, 169–70; Iranian government overthrown by, 330; life expectancy in, 62, 116; methane emissions in, 118; nuclear power in, 237–38; as primary superpower, 345; rivalry between China and, 343–44; during Second World War, 335
utility, 71; prices and, 74–78

vaccinations: against coronavirus, 266; against smallpox, 12
valence, 310
valerism, 310–14; in ecological state, 321–22; modularization in, 315; political economy of, 322–30; structure of valerist state, 330–40
Valero, Francisco, 219–20
Valmy, Battle of (1792), 332
value, subjective theory of, 71
Venice (city-state), 140–41
Verizon (firm), 261
violence, in development of capitalism, 302

Volkswagen (firm), 118
Wackernagel, Mathis, 310–11
Waerness, Eric, 175
Wallerstein, Immanuel, 344–45
Walras, Léon, 71
warfare, 163–69; in Cosmopole, 350–51; industrialization and, 296; in Japanese history, 210; technological change tied to, 266
War Production Board (World War II, U.S.), 335
Washington Consensus, 345
waste, recycling of, 316
water mills, 278, 279
water supplies, subsidence in, 12
Watt, James, 277, 286–90, 295
Weber, Max, 127
Westphalia, Treaty of (1648), 331
White, Lynn, 217
White, Michael, 166
Williams, Eric, 295
wind energy, see renewable energy sources
Winner, Langdon, 218
women: age at marriage of, 274; domestic labor performed by, 36; in San communities, 306
wood: as energy source, 190–91; see also forests
wool, 283–84
Woolf, Arthur, 290
work: automation and, 253–56; energy available for, 178; mechanical, 193

Yuan dynasty (China), 209

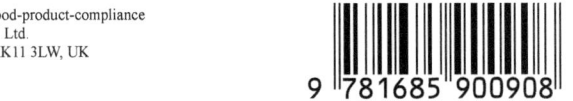

www.ingramcontent.com/pod-product-compliance
Ingram Content Group UK Ltd.
Pitfield, Milton Keynes, MK11 3LW, UK
UKHW042235060325
455777UK00005B/47